Lecture Notes in Control and Information Sciences 201

Editor: M. Thoma

Jacob Kogan

Robust Stability and Convexity

An Introduction

Springer-Verlag London Ltd.

Author

Jacob Kogan, PhD
Department of Mathematics and Statistics
University of Maryland Baltimore County
Baltimore, Maryland 21228-5398, USA

ISBN 978-3-540-19919-9 ISBN 978-3-540-39358-0 (eBook)
DOI 10.1007/978-3-540-39358-0

British Library Cataloguing in Publication Data
A catalogue record for this book is available from the British Library

Typesetting: Camera ready by author

69/3830-543210 Printed on acid-free paper

To the memory of Josef Kogan

Preface

A fundamental problem in control theory is concerned with stability of a given linear system. The system designer often wants to know if all the roots of the systems characteristic polynomial are located in a pre–specified open region of the complex plane (the left half plane, and the unit disc are important examples of such regions). In many applications the coefficients of the characteristic polynomial are functions of independent physical parameters (as, for example, coefficients of friction, spring constants, masses, capacitances, inductances, etc.). The design of a control system is generally based on a simplified model, and the true values of the physical parameters may differ from the assumed values. Consequently, it is of interest to verify root location of the entire *family* of characteristic polynomials.

When regions of stability of dynamic systems in the parameter space are described by complicated expressions it is difficult to visualize the size and the shape of these regions. An attractive way to resolve the problem is to embed a simple geometric figure inside the region. In many practical situations the polynomial family can be associated with the n dimensional box $\mathbf{B}^n = \{\mathbf{x} : 0 \leq x_i \leq 1\}$, while each frequency w is associated with a function f that maps the box into the complex plane $\mathbf{C}$. The verification of stability boils down to the following mathematical problem: Determine whether the boundary of the image of the box $\partial f(\mathbf{B}^n)$ contains the origin in the complex plane.

This problem can be reduced to a constrained optimization problem. For affine functions f the optimization problem admits an analytic solution. For nonlinear functions f there exist no analytic solutions, and associated numerical calculations are computationally difficult and time consuming.

In this manuscript we introduce and describe a set of *principal points* $\mathbf{X_p}$. The set of principal points is a subset of the box whose image under f covers the boundary of the image of the box. When f is an affine mapping in order to describe the boundary of $f(\mathbf{B}^n)$ one has to select a finite set of *principal vertices*. For example, a striking fact discovered by Kharitonov [Kh] shows that when coefficients of polynomials of degree n vary independently in pre–specified intervals there exist **four** principal vertices, and the interval polynomial family is Hurwitz stable if and only if the polynomials associated with the principal vertices are Hurwitz stable. This remarkable **four** vertices result holds **independent** of n! The set of the principal vertices is, in general, frequency dependent. Generically in the nonlinear case the set of principal points consists of finitely many *one* dimensional manifolds (which are sometimes just linear segments). As a rule the manifolds are frequency dependent, however in Section 4.4 we provide an example of a nonlinear system with frequency independent principal linear segments.

The principal points approach can be traced to the works of Zeheb and Walach [ZW], and Zeheb [Z]. Systematic applications of the approach generate new simple proofs of many known robust stability results, and lead to necessary and sufficient robust stability conditions for polynomial families with coefficients depending multiaffinely on parameters, and quasipolynomial families with uncertainties in coefficients and delays. On the other hand an application of the principal points approach to Hurwitz stability of box polynomials recovers the four vertices result (see Section 2.8).

Although the main motivation for the study of root location of characteristic functions comes from related stability problems of linear systems it is important to distinguish between the two problems. We address this issue in detail in Chapter 5.

ACKNOWLEDGMENT

I express my appreciation to Ya.Z. Tsypkin whose seminar on robust stability held in George Washington University triggered my interest in the subject. While preparing notes and reference material for a course on robust stability in the summer of 92 I was encouraged by Professor Tsypkin to integrate the material into a book.

During the last few years I have been fortunate to collaborate with N. Cohen, J. Hocherman, V.L. Kharitonov, A. Leizarowitz, B.T. Polyak, M. Teboulle, and E. Zeheb. Many of the new research results reported in this manuscript are an outgrowth of this collaboration. Many helpful suggestions and useful discussions with B.R. Barmish, F.G. Boese, N.K. Bose, M.S. Gowda, E.I. Jury, B.Ya. Kogan, D.D. Šiljak, and R. Tempo helped in building my knowledge of system theory, and many subjects related to robust stability.

My thanks also go to students at University of Maryland Baltimore County who participated in a course based on early versions of the manuscript. I would like particularly mention R. Sznajder whose remarks led to an elegant proof of Tsypkin–Polyak stability criterion given in Section 3.5.

Last but not least, I would like to express my thanks to R. Rostamian and W. Shyong for their help with computer related issues.

Contents

Chapter 1

Introduction and Motivation

No mathematical model can exactly represent the dynamics of a real physical system, so a design based on a nominal model only may not achieve the required performance criteria. The lack of precise information concerning parameters means that one cannot predict the exact output of a real engineering system even when the input is known exactly. However one can ask the question: "How much uncertainty is allowed so that the output will still satisfy given performance criteria." To be specific we consider an illustration.

1.1. Illustration

Let the model be given by

$$\dot{\mathbf{x}} = A\mathbf{x} \tag{1.1.1}$$

where $\mathbf{x}$ is an n dimensional vector, and A is an $n \times n$ matrix. The roots of the characteristic polynomial

$$\mathbf{p}(s) = a_0 + a_1 s + a_2 s^2 + \ldots + a_{n-1} s^{n-1} + s^n \tag{1.1.2}$$

determine the stability of the system (1.1.1). The system is stable if and only if the real parts of all the roots of the characteristic polynomial (1.1.2) are negative. Routh-Hurwitz criterion (see e.g., Gantmacher [Ga]) provides a way to check stability of the system.

In many applications the precise values of the coefficients a_i are not known, and one can only assume that that each coefficient a_i lies in a certain interval

$$|a_i - a_i^0| \leq \alpha_i \gamma, \qquad i = 0, \ldots, n-1. \tag{1.1.3}$$

The interval uncertainties (1.1.3) generate a box polynomial family

$$\mathbf{P}_\gamma = \left\{\mathbf{p}(s,\mathbf{a}) = \sum_{i=0}^{n-1} a_i s^i + s^n \ : \ |a_i - a_i^0| \le \alpha_i \gamma,\ i = 0,\dots,n-1\right\} \tag{1.1.4}$$

centered at the nominal polynomial

$$\mathbf{p}(s,\mathbf{a}^0) = a_0^0 + a_1^0 s + a_2^0 s^2 + \ldots + a_{n-1}^0 s^{n-1} + s^n.$$

The polynomial family $\mathbf{P}_\gamma$ is stable if every element of the family is stable, otherwise the family is unstable. This monograph is concerned with the two problems associated with families $\mathbf{P}_\gamma$.

Problem 1.1.1 Given $\gamma \ge 0$ determine whether $\mathbf{P}_\gamma$ is stable or not.

Problem 1.1.2 Given a stable nominal polynomial

$$\mathbf{p}(s,\mathbf{a}^0) = a_0^0 + a_1^0 s + a_2^0 s^2 + \ldots + a_{n-1}^0 s^{n-1} + s^n$$

find the positive scalar r (the stability radius) so that

1. $\mathbf{P}_\gamma$ is stable when $\gamma < r$.

2. $\mathbf{P}_\gamma$ is unstable when $\gamma \ge r$.

1.2. Stability and Convexity

A surprising Kharitonov's Theorem (see Theorem 2.1.1) shows that the stability of the polynomial family $\mathbf{P}_\gamma$ is determined by stability of only four special vertex polynomials. The celebrated result of Kharitonov brought a good deal of attention to the problems of robust stability in recent years and it is known, that for many important stability regions (among them the open unit disc) the stability of *all* the vertex polynomials does not necessarily guarantee the stability of the polynomial box. A remarkable result due to Rantzer [Ra1] shows that when both the stability region Ω, and its reciprocal Ω^{-1} are convex the stability of all the vertex polynomials implies the stability of the box family $\mathbf{P}_\gamma$. The stability region Ω = the open left half plane clearly satisfies Rantzer's condition, on the other hand the reciprocal of Ω = the open unit disc fails to be convex. Although stability is not a convex property, stability and convexity often appear together. An additional surprising example of the joint appearance is convexity of frequency response arcs associated with a stable polynomial discovered by Hamann and Barmish [HaB].

One hundred years ago, on September 30, 1892 A.M. Lyapunov defended his doctoral thesis on "The General Problem of the Stability of Motion". A little earlier, on March 30, 1892 Stefan Banach was born. This historic coincidence may provide a partial explanation for many important links between stability and convexity. The connection between stability and convexity is the main subject of this manuscript. Convexity techniques systematically recover many already known robust stability results, and lead to computationally tractable criteria for nonlinear problems. Necessary and sufficient robust stability conditions for particular examples of multiaffine problems and time delay systems with interval uncertainties in coefficients and delays are considered in the book.

The manuscript is based on a one semester introductory course devoted to robust stability problems given at University of Maryland Baltimore County in the Fall of 92. The course introduces a mixed population of advanced undergraduate and graduate students to basic results related to this area of research. A special effort has been made in order to keep the level of exposition as simple as possible. In selecting the material the author has been following his own research interests. Time limitations of one semester course do not allow to cover many important related topics. On the other hand the author believes in the following results due to G. Leitmann [Lei]:

Theorem. There does not exist a *best* method, that is, one which is *superior* to all other methods, for solving all problems in a given class of problems.
Proof. By contradiction.
Corollary. Those who believe, or claim, that their method is the *best* one suffer from that alliterative affliction, ignorance/arrogance.
Proof. By observation.

An attempt therefore has also been made to direct an interested reader to recent publications related to important material not covered in the manuscript. In particular I would like to mention the following sources:

- Ackermann J., Robust Control: Systems with Uncertain Physical Parameters. Springer–Verlag, London, 1993.

- Barmish B.R., New Tools for Robustness of Linear Systems. Macmillan, New York, 1994.

- Bhattacharyya S.P., Robust Stabilization Against Structural Perturbations. New York, Springer–Verlag, 1987.

- Boyd S.P., Barratt C.H., Linear Controller Design. Prentice Hall, New Jersey, 1991.

- Boyd S.P., El Ghaoui L., Feron E., Balakrishnan V., Linear Matrix Inequalities in Systems and Control Theory. SIAM Studies in Applied Mathematics, Vol. 15, 1994.

- Doyle J.C., Francis B.A., Tannenbaum A.R., Feedback Control Theory. Macmillan, New York, 1992.

- Eslami M., Theory of Sensitivity in Dynamic Systems. Springer–Verlag, New York, 1995.

- Weinmann A., Uncertain Models and Robust Control. Springer–Verlag, 1991.

- Control of Uncertain Dynamic Systems. International Workshop on Robust Control, San Antonio, Texas, Bhattacharyya S.P., Keel L.H. (eds). CRC Press, 1991.

- The Modeling of Uncertainty in Control Systems. Proceedings of the 1992 Santa Barbara Workshop, Smith R. and Dahleh M. (eds). Lecture Notes in Control and Information Sciences, Volume 192, Springer-Verlag, London, 1994.

- Special Issue on Robust Control. Automatica, Vol. 29, No. 1, 1993.

- $50th$ Anniversary Issue. Journal of Dynamic Systems, Measurment and Control, Vol. 115, No. 2(B), 1993.

- Special Issue: Horowitz and QFT Design Methods. International Journal of Robust and Nonlinear Control, Vol. 4, No. 1, 1994.

- Special Issue: Implicit and Robust Systems. Circuits, Systems and Signal Processing. Vol. 13, No. 2/3, 1994.

- Special Issue on Robustness of Multidimensional Systems. Multidimensional Systems and Signal Processing, Vol. 5, No. 4, 1994.

Furthermore, in addition to the references given at the end of the book, the reader may consult an electronic reference data base on robust stability. To access the list login into a machine which understands "ftp", and follow the instruction:

#	*you type*	*machine responses*
1	ftp 130.85.145.10	username:
2	anonymous	password:
3	kogan	ftp>
4	cd pub/kogan	ftp>
5	get refs.tex newfile	ftp>
6	quit	Goodbye

The procedure copies the LaTeX file refs.tex into your directory. The name of your new file is newfile.

Corrections, comments and suggestions for improvement are welcome!

Chapter 2

Stability of Box Polynomial Families

This chapter investigates stability of a given family of polynomials. Without loss of generality we set $\gamma = 1$, and for simplicity of notations denote the family just by $\mathbf{P}$, i.e.,

$$\mathbf{P} = \left\{ \mathbf{p}(s, \mathbf{a}) = \sum_{i=0}^{n-1} a_i s^i + s^n \ : \ a_i^0 - \alpha_i \leq a_i \leq a_i^0 + \alpha_i, \ i = 0, \ldots, n-1 \right\}.$$

In the next section we provide necessary and sufficient conditions for Hurwitz stability of the family $\mathbf{P}$.

2.1. Kharitonov's Theorem

In this section we present a surprising result concerning stability of interval polynomials.

Theorem 2.1.1 (Kharitonov [Kh].) The polynomial family $\mathbf{P}$ is stable if and only if the four polynomials

$$\mathbf{p}(s, \mathbf{a}^1) = [a_0^0 + \alpha_0] + [a_1^0 + \alpha_1]s + [a_2^0 - \alpha_2]s^2 + [a_3^0 - \alpha_3]s^3 + \ldots,$$

$$\mathbf{p}(s, \mathbf{a}^2) = [a_0^0 + \alpha_0] + [a_1^0 - \alpha_1]s + [a_2^0 - \alpha_2]s^2 + [a_3^0 + \alpha_3]s^3 + \ldots,$$

$$\mathbf{p}(s, \mathbf{a}^3) = [a_0^0 - \alpha_0] + [a_1^0 - \alpha_1]s + [a_2^0 + \alpha_2]s^2 + [a_3^0 + \alpha_3]s^3 + \ldots,$$

$$\mathbf{p}(s, \mathbf{a}^4) = [a_0^0 - \alpha_0] + [a_1^0 + \alpha_1]s + [a_2^0 + \alpha_2]s^2 + [a_3^0 - \alpha_3]s^3 + \ldots,$$

are stable.

Our proof is borrowed from [MAD] . We first remind the reader some simple and useful properties of stable polynomials.

Lemma 2.1.1 *Let* $\mathbf{p}(s) = a_0 + a_1 s + a_2 s^2 + \ldots + a_{n-1}s^{n-1} + s^n$ *be a stable polynomial. Then* $a_0 > 0$.

Proof: Suppose that $\mathbf{p}(0) = a_0 \leq 0$. Since $\mathbf{p}(s) > 0$ for large real positive s, there exists real $s^* \geq 0$ so that $\mathbf{p}(s^*) = 0$. This contradiction completes the proof.

In fact all the coefficients of the stable polynomial $\mathbf{p}(s)$ are positive (see e.g., Gantmacher [Ga]).

Lemma 2.1.2 *Suppose that* $\mathbf{p}(s)$ *is a stable polynomial. Then arg* $\mathbf{p}(jw)$ *is a continuous and strictly increasing function of* w.

Proof: The statement is obvious for polynomials of degree 1. A stable polynomial of degree n is a product of n stable polynomials of degree 1.

Lemma 2.1.3 *Let* $\mathbf{p}(s, \mathbf{a}')$ *be a stable polynomial, and* $\mathbf{p}(s, \mathbf{a}'')$ *be an unstable polynomial. There exist* $\lambda \in [0,1]$, *and a real* $w \geq 0$ *so that* $\mathbf{p}(jw, \lambda\mathbf{a}' + (1-\lambda)\mathbf{a}'') = 0$.

Proof: The roots of a monic polynomial are continuous functions of its coefficients. All the roots of the stable polynomial $\mathbf{p}(s, \mathbf{a}')$ are located in the left half plane, at least one root of the unstable polynomial $\mathbf{p}(s, \mathbf{a}'')$ lies in the right half plane. When λ varies from 0 to 1 the unstable root of $\mathbf{p}(s, \mathbf{a}'')$ travels to the left half plane. For some $\lambda \in [0,1]$ the root crosses the imaginary axis–the boundary of the stability region. That is

$$\mathbf{p}(jw, \lambda\mathbf{a}' + (1-\lambda)\mathbf{a}'') = 0.$$

Since $\mathbf{p}(s, \lambda\mathbf{a}' + (1-\lambda)\mathbf{a}'')$ is a polynomial with real coefficients one has

$$\mathbf{p}(-jw, \lambda\mathbf{a}' + (1-\lambda)\mathbf{a}'') = \overline{\mathbf{p}(jw, \lambda\mathbf{a}' + (1-\lambda)\mathbf{a}'')} = 0.$$

This completes the proof.

If at least one polynomial in $\mathbf{P}$ is stable in order to verify stability of the entire family one has to check that for each $w \geq 0$

$$0 \notin \{\mathbf{p}(jw, \mathbf{a}) \ : \ \mathbf{p}(s, \mathbf{a}) \in \mathbf{P}\}.$$

This condition has a special name "zero exclusion" criterion (see e.g., [FD]). The criterion plays an important role in robustness analysis. We will take a closer look at the set of values of the polynomial family at jw.

Consider the n dimensional box

$$\mathbf{B} = \{\mathbf{a} \ : \ \mathbf{a} \in \mathbf{R}^n, \ \underline{a}_i \leq a_i \leq \overline{a}_i, \ \ i = 0, \ldots, n-1\}$$

where $\underline{a}_i = a_i^0 - \alpha_i$, and $\overline{a}_i = a_i^0 + \alpha_i$.

Definition 2.1.1 *The value set of the polynomial family* **P** *at w is a subset $\mathcal{P}_w$ of the complex plane defined by the relation:*

$$\mathcal{P}_w = \{\mathbf{p}(jw, \mathbf{a}) \ : \ \mathbf{a} \in \mathbf{B}\}.$$

For a fixed w the mapping $\mathbf{a} \to \mathbf{p}(jw, \mathbf{a})$ is an affine mapping from $\mathbf{R}^n$ to the complex plane $\mathbf{C}$. The image of the box $\mathbf{B}$ is a polygon in the complex plane. We are ready now to prove the theorem. Let

$$\mathbf{p}(jw, \mathbf{a}) = U(w, \mathbf{a}) + jV(w, \mathbf{a})$$

where

$$U(w, \mathbf{a}) = a_0 - a_2 w^2 + a_4 w^4 - \ldots, \text{ and } V(w, \mathbf{a}) = w\left[a_1 - a_3 w^2 + a_5 w^4 - \ldots\right].$$

Since $U(w, \mathbf{a})$ is defined by even coordinates of $\mathbf{a}$, and $V(w, \mathbf{a})$ is defined by odd coordinates of $\mathbf{a}$, the value set $\mathcal{P}_w$ is a rectangle whose extreme points are generated by the Kharitonov polynomials.

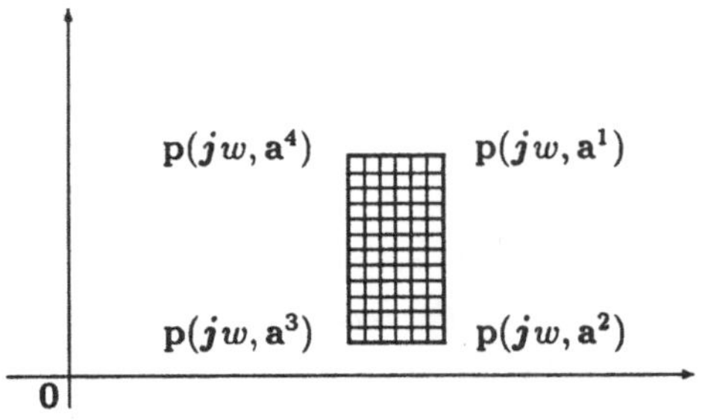

Note, that $\mathcal{P}_0$ is an interval $[a_0 - \alpha_0, a_0 + \alpha_0]$. Since Kharitonov polynomials are stable one has $a_0 - \alpha_0 > 0$, and $0 \notin \mathcal{P}_0$. As w increases from 0 to ∞ the rectangle $\mathcal{P}_w$ is rotating counterclockwise in the complex plane. In order to check the stability of the polynomial family we have to check that the value sets do not hit the origin. Suppose, for example, that $\mathcal{P}_w$ hits the origin by an interior point of its "bottom" side.

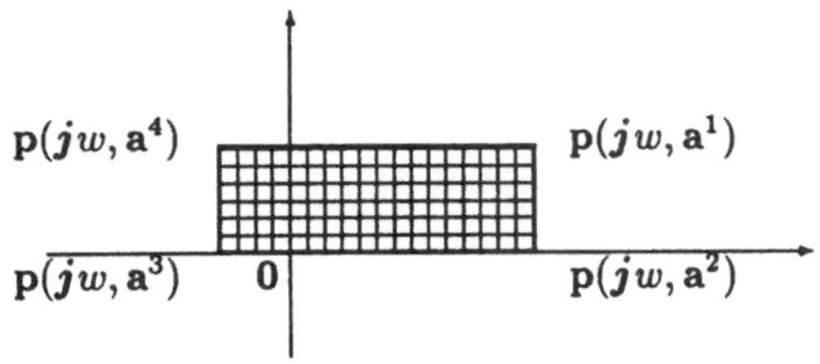

When w increases the arguments of $\mathbf{p}(jw, \mathbf{a}^2)$ and $\mathbf{p}(jw, \mathbf{a}^3)$ will increase, hence the line connecting these two points will not be parallel to the real axis anymore. This contradiction shows that the value set may hit the origin only by its extreme points. Since the extreme points are generated by the stable Kharitonov polynomials one has $\mathbf{p}(jw, \mathbf{a}^i) \neq 0$. This completes the proof.

When the degree of the box polynomial family is less than 6 stability of less than 4 vertex polynomials guarantees stability of the whole family. Specifically, for polynomials of degree 3, 4 and 5 one has to check stability of 1, 2, or 3 polynomials respectively (see [AJM], [MAD]) .

2.2. Other root location regions: Ω stability

In many cases the stability of the system is maintained when the roots of the characteristic polynomial are contained in regions other then the left half plane. When, for example, one is interested in stability of a discrete time system, the stability region is the unit disc. An additional important class of stability regions are sectors in the complex plane.

Definition 2.2.1 *Let Ω be an open subset of the complex plane. A polynomial $\mathbf{p}(s)$ is Ω-stable if and only if all the roots of the polynomial lie in Ω. We shall say that the polynomial family $\mathbf{P}$ is Ω-stable if and only if every polynomial in $\mathbf{P}$ is Ω-stable.*

Kharitonov's Theorem naturally motivates the following questions:

1. Under what condition "extreme point" results can be extended for root location regions other than the left–half–plane?

2. How robust stability problems can be handled when "extreme point" results do not hold?

Bose and Zeheb [BZ] investigated Schur stability (i.e., Ω is the unit disc centered at the origin) of the interval polynomial family

$$\mathbf{p}(s, \mathbf{a}) = -\frac{1}{3} + \frac{3}{2}s^2 + a_3 s^3 + s^4, \qquad -\frac{17}{8} \leq a_3 \leq \frac{17}{8}. \tag{2.2.1}$$

Although all the "vertex" polynomials of the family are Ω stable, the polynomial $-\frac{1}{3} + \frac{3}{2}s^2 + s^4$ has two roots $s = \pm j1.3025$ which lie outside the unit disc.

On the other hand Foo and Soh [FS1] investigated "sector" stability, i.e.,

$$\Omega = \left\{ s \ : \ s = re^{j\phi}, \ |\phi - \pi| \leq \phi_0, \ r \geq 0 \right\}, \qquad \phi_0 < \frac{\pi}{2}$$

of an interval polynomial family. It turns out that the family is sector stable if and only if $2n$ special vertex polynomials are stable. A recipe for construction the vertex polynomials is provided in the paper.

As we can see vertex polynomials alone provide crucial information for some stability regions. Stability regions Ω for which stability of vertex polynomials implies stability of the interval polynomial family for *any* box polynomial family are called *weak Kharitonov* regions. Petersen [Pe] discovered an important class of weak Kharitonov regions, and also showed that intersection of two weak Kharitonov regions is again a weak Kharitonov region. A general characterization of weak Kharitonov regions is due to Rantzer [Ra1]:

Theorem 2.2.1 Ω is a weak Kharitonov region if Ω and $\Omega^{-1} = \left\{ z \ : \ z^{-1} \in \Omega \right\}$ are convex.

In the case of box polynomials with complex coefficients Rantzer's result provides necessary and sufficient conditions for weak Kharitonov regions. The theorem clearly indicates that "extreme point" results do not hold in general when one is interested in stability of discrete time systems. In the next section we tackle Schur stability problem for interval families.

2.3. Principal Vertices and Stability

Although this section is motivated by stability criteria for interval families and a specific root location domain–the unit disc, we first set up a slightly more general framework.

Definition 2.3.1 *The unit box in* $\mathbf{R}^n$ *is* $\mathbf{B}^n = \{\mathbf{x} \ : \ \mathbf{x} \in \mathbf{R}^n, \ 0 \leq x_i \leq 1\}$. *The set of the vertices of the box is denoted by* $\mathbf{V}^n$.

When it does not lead to ambiguity we drop the superscript n, and denote the box, and the set of vertices by $\mathbf{B}$ and $\mathbf{V}$ respectively.

For a given root location domain Ω let $\delta \ : \ I_\Omega \rightarrow \partial\Omega$ be a parameterization of the boundary $\partial\Omega$ of the domain (i.e., when, for example, Ω is the open left half plane, then $I_\Omega = (-\infty, \infty)$, and $\delta(w) = jw$; when Ω is the open unit disc, then $I_\Omega = [0, 2\pi)$, and $\delta(w) = e^{jw}$). Consider a real interval polynomial family

$$\mathbf{P} = \{\mathbf{p}(s, \mathbf{x}) \; : \; \mathbf{p}(s, \mathbf{x}) = a_0(\mathbf{x}) + a_1(\mathbf{x})s + \ldots + a_{n-1}(\mathbf{x})s^{n-1} + s^n, \; \mathbf{x} \in \mathbf{B}^n\}, \quad (2.3.1)$$

where $a_i(x) = \underline{a}_i + \alpha_i x_{i+1}$, $\alpha_i > 0$, $i = 0, \ldots, n-1$, $n \geq 2$.

A general stability result covering polytopes of polynomials and domains Ω is the Edge Theorem of Bartlett, Hollot and Huang [BHH]. The Edge Theorem tells as the following: To verify Ω stability of a polytope of polynomials one has to check Ω stability of the edges of the polytope. Since the edges are only one dimensional we obtain a drastic reduction in computational complexity. The number of the edges, however, may be large. Our goal in this section is to provide a stability criterion that further reduces the computational effort.

If for some $\mathbf{x}^0 \in \mathbf{B}^n$ the polynomial $\mathbf{p}(s, \mathbf{x}^0)$ is Ω stable, then the "zero exclusion" criterion implies that the family $\mathbf{P}$ is Ω stable if and only if

$$0 \notin \mathcal{P}_w \text{ for each } w \in I_\Omega. \quad (2.3.2)$$

In fact for polynomials with real coefficients I_Ω can be substituted by

$$I_\Omega^+ = \{w \in I_\Omega, \; \text{Im}\, \delta(w) \geq 0\}. \quad (2.3.3)$$

Since $\mathbf{x} \to \mathbf{p}(s, \mathbf{x})$ is an affine mapping for each $s \in \mathbf{C}$, the value set $\mathcal{P}_w$ is a polygon for each $w \in I_\Omega$. When one considers a polygon in the complex plane whose consecutive vertices are $\{z_1, \ldots, z_N\}$, $N \geq 3$, then a point z in the complex plane does not belong to the polygon if and only if

$$\min_{1 \leq i \leq N} \left\{ \text{Im} \, \frac{z - z_i}{z_{i+1} - z_i} \right\} < 0, \quad (2.3.4)$$

where $z_{N+1} = z_1$. Suppose that the consecutive vertices of the polygon $\mathcal{P}_w$ are generated by the vertices

$$\left\{ \mathbf{v}_p^1(w), \ldots, \mathbf{v}_p^{N(w)}(w) \right\}. \quad (2.3.5)$$

The stability criterion given below follows from (2.3.2) and (2.3.4).

Theorem 2.3.1 The polynomial family $\mathbf{P}$ is Ω stable if and only if

1. There exists $\mathbf{x}^0 \in \mathbf{B}^n$ so that $\mathbf{p}(s, \mathbf{x}^0)$ is Ω stable.

2. For each $w \in I_\Omega^+$ one has $\mathbf{p}(\delta(w), \mathbf{v}_p^1(w)) \neq 0$, and

$$\begin{cases} \text{if } N(w) = 2, \text{ then } \operatorname{Im} \frac{\mathbf{p}(\delta(w),\mathbf{v}_p^2(w))}{\mathbf{p}(\delta(w),\mathbf{v}_p^1(w))} \neq 0 \\ \text{if } N(w) = 2 \text{ and } \operatorname{Im} \frac{\mathbf{p}(\delta(w),\mathbf{v}_p^2(w))}{\mathbf{p}(\delta(w),\mathbf{v}_p^1(w))} = 0, \text{ then } \operatorname{Re} \frac{\mathbf{p}(\delta(w),\mathbf{v}_p^2(w))}{\mathbf{p}(\delta(w),\mathbf{v}_p^1(w))} > 0 \\ \text{if } N(w) \geq 3, \text{then } \min\limits_{1 \leq i \leq N(w)} \left\{ \frac{-\mathbf{p}(\delta(w),\mathbf{v}_p^i(w))}{\mathbf{p}(\delta(w),\mathbf{v}_p^{i+1}(w)) - \mathbf{p}(\delta(w),\mathbf{v}_p^i(w))} \right\} < 0. \end{cases}$$

It is of interest to identify a polynomial subfamily $\mathbf{P_T}$ of $\mathbf{P}$ so that the family $\mathbf{P}$ is Ω stable when $\mathbf{P_T}$ is Ω stable. The subfamily $\mathbf{P_T}$ will be called a testing family for $\mathbf{P}$. The Edge Theorem provides an example of a testing family. We next consider an example of a polynomial subfamily $\mathbf{P_T}$ smaller than the set of exposed edges.

Let $\mathbf{P_T}$ be a subfamily of $\mathbf{P}$ defined as follows

$$\mathbf{P_T} = \left\{ \lambda \mathbf{p}(s, \mathbf{v}_p^i(w)) + (1-\lambda)\mathbf{p}(s, \mathbf{v}_p^{i+1}(w)) \; : \; \begin{array}{c} \lambda \in [0,1] \\ i = 1, \ldots, N(w) \\ w \in I_\Omega^+ \end{array} \right\}. \tag{2.3.6}$$

An alternative stability condition is given next.

Theorem 2.3.2 The polynomial family $\mathbf{P}$ is Ω stable if and only if the subfamily $\mathbf{P_T}$ is Ω stable.

According to the Edge Theorem, the number of polynomial segments required to be checked in order to verify Schur stability of an interval polynomial family is $n2^{n-1}$. On the other hand, as we will see later, the number of polynomial segments required to be checked by the "principal vertices" technique does not exceed $2n^3$. In many cases the second condition of the theorem can be checked straightforward.

Our next goal is to characterize the function $N(w)$, and to investigate the dynamics of the set (2.3.5).

2.4. Principal Directions and Principal Vertices

Let $w \in I_\Omega$ be fixed. Consider the image of the box $\mathbf{B}$ under the affine transformation

$$f(\mathbf{x}) = \mathbf{p}(\delta(w), \mathbf{x}).$$

Definition 2.4.1 *Let* $\mathbf{v} \in \mathbf{V}$. *A vertex* $\mathbf{w}$ *is a neighboring vertex if there exists an index* i *such that*

$$v_i \neq w_i, \text{ and } v_k = w_k \text{ when } k \neq i.$$

The set of neighboring vertices is denoted by $\mathbf{N_v}$. *The half line*

$$\alpha[f(\mathbf{w}) - f(\mathbf{v})], \; \mathbf{w} \in \mathbf{N_v}, \; f(\mathbf{w}) \neq f(\mathbf{v}), \; \alpha > 0 \tag{2.4.1}$$

is an edge direction at $f(\mathbf{v})$. *The set of edge directions at* $f(\mathbf{v})$ *is denoted by* $\mathbf{E_v}$.

Lemma 2.4.1 *Let* $\mathbf{v} \in \mathbf{V}$. *If there exists an open half plane that contains* $\mathbf{E_v}$, *then* $f(\mathbf{v})$ *is an extreme point of* $f(\mathbf{B})$.

Proof: We proof the lemma first for the special case $\mathbf{v} = (0, \ldots, 0)^t$. Let $\mathbf{v}^i \in \mathbf{N_v}$, i.e.,

$$\mathbf{v}^i = \left(v_1^i, \ldots, v_n^i\right)^t \text{ so that } v_i^i = 1, \text{ and } v_k^i = 0 \text{ when } k \neq i.$$

Let $\mathbf{x} \neq \mathbf{v}$ be an element of the box, then

$$\mathbf{x} = \sum_{i=1}^{n} x_i \mathbf{v}^i, \qquad x_i \in [0, 1], \qquad \sum_{i=1}^{n} x_i > 0.$$

Since f is affine, and $\mathbf{v} = (0, \ldots, 0)^t$ one has

$$f(\mathbf{x}) - f(\mathbf{v}) = f\left(\sum_{i=1}^{n} x_i \mathbf{v}^i\right) - f(\mathbf{v}) = \sum_{i=1}^{n} x_i \left[f(\mathbf{v}^i) - f(\mathbf{v})\right].$$

This shows that either $f(\mathbf{x}) - f(\mathbf{v}) = 0$, or $f(\mathbf{x}) - f(\mathbf{v})$ belongs to the open half plane that contains $\mathbf{E_v}$. Hence $f(\mathbf{v})$ is an extreme point of $f(\mathbf{B})$. The proof of the lemma for an arbitrary vertex $\mathbf{v}$ is reduced to the special case by applying the mapping $\mathbf{x} \to \mathbf{x} - \mathbf{v}$, and observing that affine transformations preserve convexity.

The lemma justifies the following definition.

Definition 2.4.2 *A vertex* $\mathbf{v}$ *is a principal vertex if and only if there exists an open half plane that contains* $\mathbf{E_v}$.

Let

$$G_1 = \frac{\partial f}{\partial x_1}, \ldots, G_n = \frac{\partial f}{\partial x_n}.$$

Since at each w, the function $f(\mathbf{x}) = \mathbf{p}(\delta(w), \mathbf{x})$ is affine with respect to $\mathbf{x}$, the partial derivatives G_i do not depend on $\mathbf{x}$, and are functions of w only, i.e., $G_i = G_i(w)$. If $\mathbf{w} \in \mathbf{N_v}$, then $f(\mathbf{w}) - f(\mathbf{v})$ is $\pm G_i(w)$. This simple observation motivates the definition of principal directions.

Definition 2.4.3 *The half lines generated by the complex numbers*

$$\pm G_1(w), \ldots, \pm G_n(w) \tag{2.4.2}$$

are principal directions at w.

In order to characterize extreme points of the value set we need to introduce additional definitions.

Definition 2.4.4 *Let z_1 and z_2 be nonzero complex numbers. We say that*

$$z_1 \prec z_2 \text{ if } 0 < Im\,\frac{z_2}{z_1},\quad z_1 \preceq z_2 \text{ if } 0 \leq Im\,\frac{z_2}{z_1},\quad \text{and } z_1 \asymp z_2 \text{ if } 0 = Im\,\frac{z_2}{z_1}.$$

We consider nonzero elements of (2.4.2) generating distinct directions. The elements of this set listed in the order of increasing arguments are denoted by

$$D_1(w) \prec \ldots \prec D_{2n(w)}(w), \qquad D_{n(w)+i} = -D_i. \tag{2.4.3}$$

The number of principal directions is $0 \leq 2n(w) \leq 2n$ (some of the numbers $\pm G_i$ may vanish, others may be proportional). When it does not lead to ambiguity we simplify the notations, and drop the argument w.

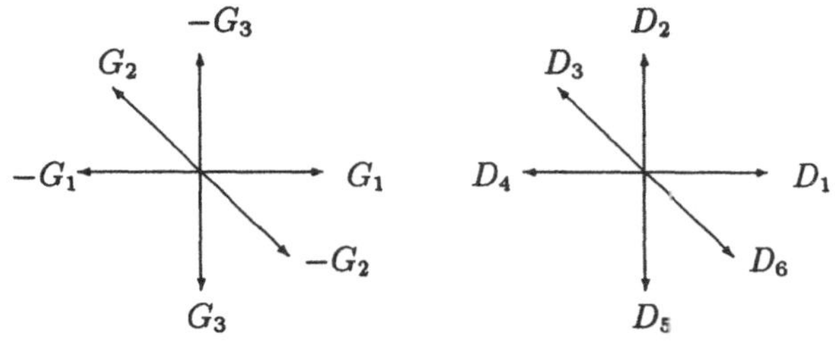

Figure 1: Wheel diagram, 6 principal directions at w.

Let $\mathbf{v} = (v_1, \ldots, v_n)^t \in \mathbf{V}$. The edge directions $\mathbf{E_v}$ are generated by the complex numbers

$$(-1)^{v_1}G_1,\ \ldots,\ (-1)^{v_n}G_n. \tag{2.4.4}$$

To identify a principal vertex one has to select $n(w)$ half lines from $2n(w)$ principal directions in the wheel diagram (see Figure 1). The selected set of the half lines should be contained in an open half plane. A half plane in the complex plane is defined by a non zero complex number. The vertex $\mathbf{v}$ is a principal vertex if and only if there exists a non zero complex number G such that

$$(-1)^{v_i}G_i(\mathbf{v}) \prec G \text{ for each } G_i \neq 0.$$

In other words $\mathbf{E_v}$ must be generated by exactly one of the sets listed below:

$$\begin{array}{lll} \{D_1 & , \ \ldots, & D_{n(w)}\} \\ \{D_2 & , \ \ldots, & D_{n(w)+1}\} \\ \ldots & , \ \ldots, & \ldots \\ \{D_{2n(w)}, & \ldots, & D_{n(w)-1}\}. \end{array} \tag{2.4.5}$$

The next statement summarizes this discussion.

Lemma 2.4.2 *There exists a one to one correspondence between the extreme points of the value set $\mathcal{P}_w$, and the $2n(w)$ elements of (2.4.5).*

Proof: Let z is an extreme point of the value set, and $\mathbf{v} \in \mathbf{V}$ so that $z = f(\mathbf{v})$. If there exists a vertex $\mathbf{w} \neq \mathbf{v}$ such that $z = f(\mathbf{w})$, then

$$\{(-1)^{v_1}G_1, \ldots, (-1)^{v_n}G_n\} = \{(-1)^{w_1}G_1, \ldots, (-1)^{w_n}G_n\}.$$

This shows that the extreme point z can be associated with a unique set

$$\left\{D_k, D_{k+1}, D_{k+2}, \ldots, D_{k+n(w)-1}\right\}. \tag{2.4.6}$$

On the other hand consider a set (2.4.6), and a vertex $\mathbf{v} = (v_1, \ldots, v_n)^t$ defined by

$$v_i = \begin{cases} 0 & \text{if} \quad G_i \in \{D_k, \ldots, D_{k+n(w)-1}\}, \\ 1 & \text{if} \ -G_i \in \{D_k, \ldots, D_{k+n(w)-1}\}, \\ 0 & \text{otherwise.} \end{cases}$$

The vertex $\mathbf{v}$ is a principal vertex, and the extreme point $z = f(\mathbf{v})$ is associated with (2.4.6).

Finally we present an algorithm that selects a subset of principal vertices generating extreme points of $\mathcal{P}_w$. Let $\mathbf{V}_p$ be the set of principal vertices of the box $\mathbf{B}$ under f. With each principal vertex $\mathbf{v}$ we associate the "left" vertex $\mathbf{v}^l$, and the "right" vertex $\mathbf{v}^r$ such that:

1. for each vertex $\mathbf{w}$ one has $f(\mathbf{v}^r) - f(\mathbf{v}) \preceq f(\mathbf{w}) - f(\mathbf{v}) \preceq f(\mathbf{v}^l) - f(\mathbf{v})$,
2. if $f(\mathbf{w}) - f(\mathbf{v}) = \alpha[f(\mathbf{v}^r) - f(\mathbf{v})]$, then $0 \leq \alpha \leq 1$,
3. if $f(\mathbf{w}) - f(\mathbf{v}) = \alpha[f(\mathbf{v}^l) - f(\mathbf{v})]$, then $0 \leq \alpha \leq 1$.

It is easy to show that a subset of $\mathbf{V}_p$ whose images are vertices of the polygon $f(\mathbf{B})$ can be generated as follows.

Algorithm 2.4.1 (Selection of a minimal subset of principal vertices.)
Step 1. Select any $\mathbf{v}^1 \in \mathbf{V}_p$.
Step 2. If $\mathbf{v}^k \in \mathbf{V}_p$ has already been selected, select $\mathbf{v}^{k+1} = \left(\mathbf{v}^k\right)^r$.
Step 3. If $\mathbf{v}^{k+1} = \mathbf{v}^1$ then **stop**, else goto **step 2**.

Remark 2.4.1 It is easy to see that when $\delta(w) \neq 0$ there is a one–to–one correson-dence between the principal vertices and the extreme points of the polygon $f(\mathbf{B})$ (see e.g., [LCZ]). When $\delta(w) = 0$ the value set $f(\mathbf{B})$ is a real interval and Algorithm 2.4.1 selects two principal vertices whose images cover the end points of the interval.

We complete the section with a formula for $\mathbf{p}(\delta(w),\mathbf{v}^{i+1})-\mathbf{p}(\delta(w),\mathbf{v}^{i})$, where $\mathbf{v}^1,\mathbf{v}^2,\dots$ is a finite sequence of principal vertices generated by Algorithm 2.4.1. The formula is instrumental in deriving extreme point results in Section 2.8. To simplify the exposition we select $\mathbf{v}^1$ that corresponds to the set $\left\{D_1,\dots,D_{n(w)}\right\}$.

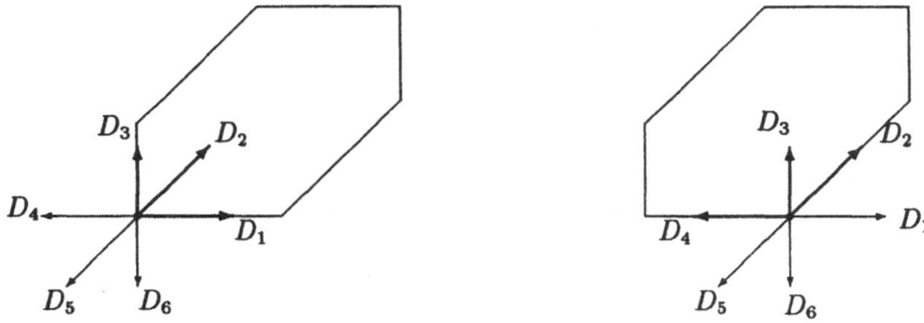

$f(\mathbf{v}^1)$ and D_1, D_2, D_3 $\qquad$ $f(\mathbf{v}^2)$ and D_2, D_3, $-D_1$

If $\mathbf{v}^1$ corresponds to $\left\{D_1,\dots,D_{n(w)}\right\}$, then

$\mathbf{v}^i$ corresponds to $\left\{D_i,D_{i+1},\dots,D_{n(w)+i-1}\right\}$,

and

$\mathbf{v}^{i+1}$ corresponds to $\left\{D_{i+1},\dots,D_{n(w)+i-1},D_{n(w)+i}\right\}=\left\{D_{i+1},\dots,D_{n(w)+i-1},-D_i\right\}$.

Let $\mathbf{v}^i=(e_1,\dots,e_n)^t$, where e_m, $m=1,\dots,n$ is an "extreme" point of the interval $[0,1]$. Let $i_1,\dots,i_k$ be indices so that the partial derivatives

$$(-1)^{e_{i_1}}\frac{\partial f}{\partial x_{i_1}},\dots,(-1)^{e_{i_k}}\frac{\partial f}{\partial x_{i_k}},\ \text{generate the } i\text{th half line } \{\beta D_i\ :\ \beta>0\},$$

i.e.,

$$(-1)^{e_{i_1}}\frac{\partial f}{\partial x_{i_1}}=\beta_1 D_i,\dots,(-1)^{e_{i_k}}\frac{\partial f}{\partial x_{i_k}}=\beta_k D_i,\qquad \beta_1,\dots,\beta_k>0. \tag{2.4.7}$$

The coordinates of $\mathbf{v}^{i+1}=(e'_1,\dots,e'_n)^t$ are given by

$$e'_m=\begin{cases} e_m & \text{if } m\notin\{i_1,\dots,i_k\}\\ 1-e_m & \text{if } m\in\{i_1,\dots,i_k\}\end{cases}$$

Hence

$$\begin{aligned}\mathbf{p}(\delta(w),\mathbf{v}^{i+1})-\mathbf{p}(\delta(w),\mathbf{v}^{i}) &= 2\left[(-1)^{e'_{i_1}}\frac{\partial f}{\partial x_{i_1}}+\dots+(-1)^{e'_{i_k}}\frac{\partial f}{\partial x_{i_k}}\right]\\ &= -2\left[\beta_1+\dots+\beta_k\right]D_i(w).\end{aligned}$$

When $\delta(w) = we^{\mathbf{j}\phi}$ the complex number $D_i(w)$ is proportional to $e^{m\mathbf{j}\phi}$, i.e., the argument of $D_i(w)$ does not depend on w, and

$$\frac{d}{dw} \arg \left[\mathbf{p}(\delta(w), \mathbf{v}^{i+1}) - \mathbf{p}(\delta(w), \mathbf{v}^{i})\right] = 0. \tag{2.4.8}$$

In Section 2.8 we show that equation (2.4.8) leads to extreme point results discovered by Foo and Soh [FS1].

2.5. Principal Vertices and Testing Family

The number of principal directions $2n(w)$ may be less than $2n$. The principal directions are generated by the partial derivatives $\frac{\partial f}{\partial x_k} = \alpha_{k-1}\left[\delta(w)\right]^{k-1}$. We next identify points $\theta \in [0, \pi]$ for which

$$\arg \delta(w) = \theta \quad \text{implies} \quad n(w) < n.$$

The partition $\{\theta_i\}_{i=0}^{k}$ described below was introduced by Kraus and Mansour [KM].

Lemma 2.5.1 *Let the numbers*

$$\{\theta_i\}_{i=0}^{k} = \left\{\pi \frac{l}{m}\right\},\ l = 0, 1, \ldots, n-1,\ m = l, l+1, \ldots, n-1,\ m > 0 \tag{2.5.1}$$

be written in the increasing order

$$0 = \theta_0 < \theta_1 < \ldots < \theta_k = \pi \tag{2.5.2}$$

Then

$$k = \begin{cases} 1 & \text{if } n = 2 \\ 1 + \sum_{l=2}^{n-1} \phi(l) & \text{if } n > 2 \end{cases} \tag{2.5.3}$$

where ϕ is the Euler function, i.e., $\phi(l)$ is the number of positive integers, including 1, which are less then l and are also relatively prime to l (see e.g., [AG]).

Proof: Consider the table

l	m	l/m
0	$1 \ 2 \ \ldots \ \ldots \ n-1$	$0 \ 0 \ \ldots \ \ldots \ 0 \ \ldots \ 0$
1	$1 \ 2 \ \ldots \ \ldots \ n-1$	$1 \ \frac{1}{2} \ \ldots \ \ldots \ \frac{1}{l} \ \ldots \ \frac{1}{n-1}$
2	$2 \ \ldots \ \ldots \ n-1$	$1 \ \frac{2}{3} \ \ldots \ \frac{2}{l} \ \ldots \ \frac{2}{n-1}$
l	$l \ \ldots \ n-1$	$\frac{l}{l} \ \ldots \ \frac{l}{n-1}$
$n-1$	$n-1$	$\frac{n-1}{n-1}$

We wish to count the number of the different elements in the table that contains values for l/m. Note, that the first column of the "l/m" table contains 2 different numbers 0 and 1. The numbers contained in the lth column are all different. Furthermore, they are not contained in the previous columns only if they are relatively prime to l. Hence, the contribution of the lth column is $\phi(l)$, and $k = 1 + \sum_{l=2}^{n-1} \phi(l)$.

Lemma 2.5.2 *Let $z = e^{j\theta}$, with $0 \le \theta \le \pi$. The complex numbers $\left\{\pm 1, \pm z, \ldots, \pm z^{n-1}\right\}$ generate $2n$ directions if and only if $\theta \notin \{\theta_i\}_{i=0}^{k}$.*

Proof: Suppose for a moment that z^p and z^q, $0 \le p < q \le n-1$ generate the same direction. Then $l\pi + p\theta = q\theta$, and $l \ge 0$. In other words $l\pi = m\theta$, where $m = q - p$. This implies $0 \le l \le m$, $\theta = \pi \frac{l}{m}$, and completes the proof.

Theorem 2.5.1 There exists a finite subset $\mathbf{P_T}$ of one dimensional polynomial segments of $\mathbf{P}$ such that:

1. The number of one dimensional segments in $\mathbf{P_T}$ does not exceed $2n^3$.
2. The polynomial family $\mathbf{P}$ is Ω stable if and only if the testing family $\mathbf{P_T}$ is Ω stable.

Proof:

$$\mathbf{p}(\delta(w), \mathbf{x}) = \underline{a}_0 + \alpha_0 x_1 + [\underline{a}_1 + \alpha_1 x_2]\delta(w) + \ldots + [\underline{a}_{n-1} + \alpha_{n-1} x_n][\delta(w)]^{n-1} + [\delta(w)]^n.$$

$$G_1(w) = \alpha_0, \ G_2(w) = \alpha_1 \delta(w), \ \ldots, G_n(w) = \alpha_{n-1}[\delta(w)]^{n-1}.$$

Let $\mathbf{W}$ be the set of pairs of vertices defined as follows

$$\mathbf{W} = \left\{ \left(\mathbf{v}^-, \mathbf{v}^+\right) \;:\; \mathbf{v}^-, \mathbf{v}^+ \in \mathbf{V},\ \exists w \in I_\Omega^+ \text{ such that } \begin{array}{c} \mathbf{v}^- = \mathbf{v}_p^i(w), \\ \mathbf{v}^+ = \mathbf{v}_p^{i+1}(w) \end{array} \right\}. \tag{2.5.4}$$

According to (2.4.4) and Lemma 2.4.2 when $\arg \delta(w_1) = \arg \delta(w_2)$ one has

$$\begin{array}{l} n(w_1) = n(w_2), \\ \text{and} \\ \left\{\mathbf{v}_p^1(w_1), \ldots, \mathbf{v}_p^{2n(w_1)}(w_1)\right\} = \left\{\mathbf{v}_p^1(w_2), \ldots, \mathbf{v}_p^{2n(w_2)}(w_2)\right\}. \end{array} \tag{2.5.5}$$

Furthermore, due to Lemma 2.5.2 when $\arg \delta(w_1)$, $\arg \delta(w_2) \in (\theta_i, \theta_{i+1})$, $i = 0, \ldots, k-1$, then $n(w_1) = n(w_2) = n$. A simple continuity argument shows that

$$\left\{\mathbf{v}_p^1(w_1), \ldots, \mathbf{v}_p^{2n}(w_1)\right\} = \left\{\mathbf{v}_p^1(w_2), \ldots, \mathbf{v}_p^{2n}(w_2)\right\}. \tag{2.5.6}$$

Hence the number of elements in $\mathbf{W}$ does not exceed $2n \cdot 2k$. Keeping in mind (2.5.3) we see that $k < \frac{n(n-1)}{2}$, and $2n \cdot 2k < 2n^3$.

Remark 2.5.1 The number of one dimensional segments in the testing family $\mathbf{P_T}$ does not exceed $2n^3$. When $\arg \delta(w)$ is constant the number of the segments does not exceed $2n$ (see Example 2.6.2, and Example 2.6.3).

2.6. Examples: Schur, Hurwitz, and Sector Stability

Example 2.6.1 Schur stability of real interval polynomials of degree 3.
The parameterization is $\delta(w) = e^{jw}$, and $I_\Omega^+ = [0, \pi)$. The principal directions are generated by the complex numbers

$$\{\pm G_1(w), \ldots, \pm G_n(w)\} = \left\{\pm 1, \pm e^{jw}, \pm e^{2jw}, \ldots, \pm e^{(n-1)jw}\right\}.$$

In fact the left hand side is different from the right hand side, however they generate the same directions. The corresponding wheel diagrams are given next. An application of Lemma 2.4.2 yields the principal vertices $\left\{\mathbf{v}_p^1(w), \ldots, \mathbf{v}_p^{2n(w)}(w)\right\}$, and the numbers $n(w)$.

Case 1. $0 < w < \frac{\pi}{2}$.

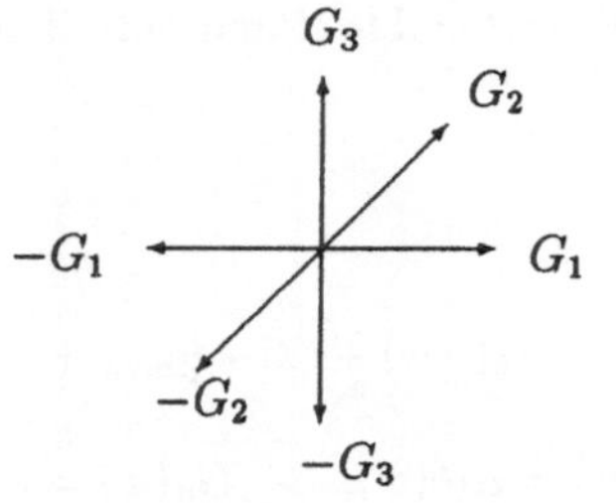

$n(w) = 3$, the six principal vertices are

$$\mathbf{v}_p^1 = \begin{bmatrix} 0 \\ 0 \\ 0 \end{bmatrix}, \ \mathbf{v}_p^2 = \begin{bmatrix} 1 \\ 0 \\ 0 \end{bmatrix}, \ \mathbf{v}_p^3 = \begin{bmatrix} 1 \\ 1 \\ 0 \end{bmatrix}, \ \mathbf{v}_p^4 = \begin{bmatrix} 1 \\ 1 \\ 1 \end{bmatrix}, \ \mathbf{v}_p^5 = \begin{bmatrix} 0 \\ 1 \\ 1 \end{bmatrix}, \ \mathbf{v}_p^6 = \begin{bmatrix} 0 \\ 0 \\ 1 \end{bmatrix}.$$

Case 2. $w = \dfrac{\pi}{2}$.

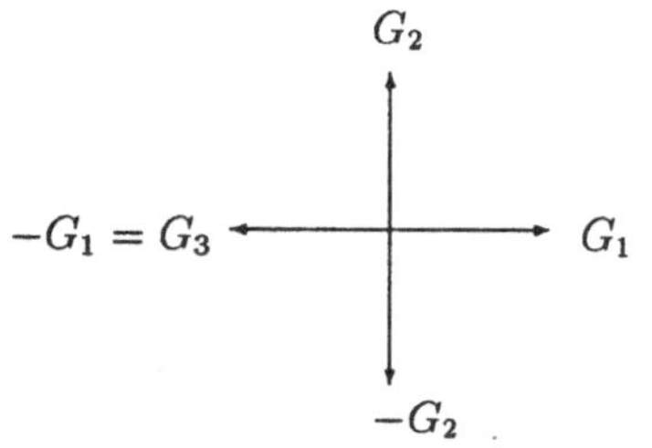

$n(w) = 2$, the four principal vertices are

$$\mathbf{v}_p^1 = \begin{bmatrix} 0 \\ 0 \\ 1 \end{bmatrix}, \ \mathbf{v}_p^2 = \begin{bmatrix} 1 \\ 0 \\ 0 \end{bmatrix}, \ \mathbf{v}_p^3 = \begin{bmatrix} 1 \\ 1 \\ 0 \end{bmatrix}, \ \mathbf{v}_p^4 = \begin{bmatrix} 0 \\ 1 \\ 1 \end{bmatrix}.$$

Case 3. $\dfrac{\pi}{2} < w < \pi$.

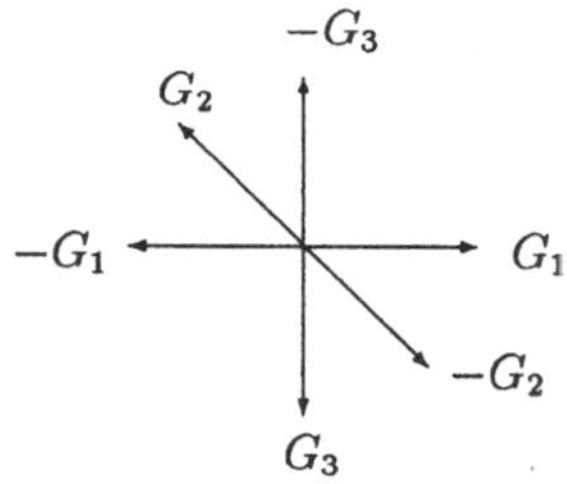

$n(w) = 3$, the six principal vertices are

$$\mathbf{v}_p^1 = \begin{bmatrix} 0 \\ 0 \\ 1 \end{bmatrix}, \ \mathbf{v}_p^2 = \begin{bmatrix} 1 \\ 0 \\ 1 \end{bmatrix}, \ \mathbf{v}_p^3 = \begin{bmatrix} 1 \\ 0 \\ 0 \end{bmatrix}, \ \mathbf{v}_p^4 = \begin{bmatrix} 1 \\ 1 \\ 0 \end{bmatrix}, \ \mathbf{v}_p^5 = \begin{bmatrix} 0 \\ 1 \\ 0 \end{bmatrix}, \ \mathbf{v}_p^6 = \begin{bmatrix} 0 \\ 1 \\ 1 \end{bmatrix}.$$

Case 4. $w = 0, \ \pi$.

$n(w) = 1$, the two principal vertices are

$$\mathbf{v}_p^1 = \begin{bmatrix} 0 \\ 0 \\ 0 \end{bmatrix}, \quad \mathbf{v}_p^2 = \begin{bmatrix} 1 \\ 1 \\ 1 \end{bmatrix}.$$

Example 2.6.2 Hurwitz stability of real interval polynomials.
The parameterization is $\delta(w) = \boldsymbol{j}w$, and $I_\Omega^+ = [0, \infty)$. The principal directions are generated by the numbers

$$\{\pm G_1(w), \ldots, \pm G_n(w)\} = \left\{\pm 1, \pm \boldsymbol{j}w, \pm(\boldsymbol{j}w)^2, \ldots, \pm(\boldsymbol{j}w)^{(n-1)}\right\}.$$

When $w > 0$ there exist exactly four principal directions, when $w = 0$ the number of principal directions is two.

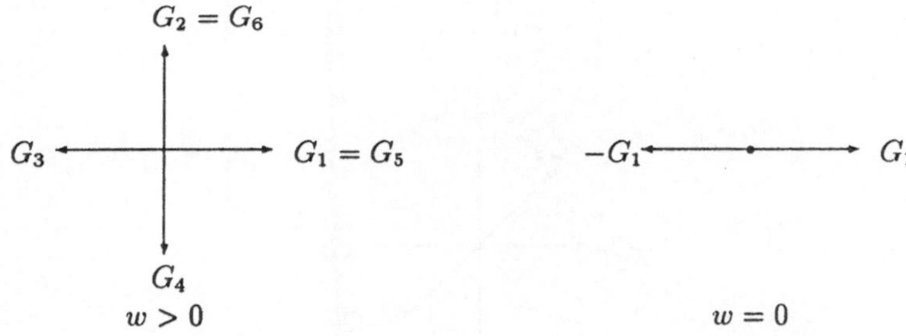

The four principal vertices $\mathbf{v}_p^1$, $\mathbf{v}_p^2$, $\mathbf{v}_p^3$, $\mathbf{v}_p^4$ corresponding to $w > 0$ are such that

Principal vertex	Edge directions are contained in
$\mathbf{v}_p^1$	the first quadrant
$\mathbf{v}_p^2$	the second quadrant
$\mathbf{v}_p^3$	the third quadrant
$\mathbf{v}_p^4$	the fourth quadrant

The four principal vertices generate the four Kharitonov polynomials . When $w = 0$ every vertex of the box is a principal vertex. The value set $\mathcal{P}_0$ is a closed interval with two extreme points $\underline{a}_0$ and $\underline{a}_0 + \alpha_0$. Algorithm 2.4.1 selects two principal vertices generating these extreme points.

Example 2.6.3 Sector stability of real interval polynomials.
The parameterization of the boundary is $\delta(w) = we^{\mathbf{j}\phi}$, and $I_\Omega^+ = [0, \infty)$. The principal directions are generated by the numbers

$$\{\pm G_1(w), \ldots, \pm G_n(w)\} = \left\{\pm 1, \pm we^{\mathbf{j}\phi}, \pm w^2 e^{2\mathbf{j}\phi}, \ldots, \pm w^{n-1} e^{(n-1)\mathbf{j}\phi}\right\}.$$

When $w > 0$ there exist at **most** $2n$ principal directions, the principal vertices do not depend on w, and coincide with the ones reported by Foo and Soh [FS1]. When $w = 0$ every vertex of the box is a principal vertex, the value set $\mathcal{P}_0$ is a closed interval with two extreme points $\underline{a}_0$ and $\underline{a}_0 + \alpha_0$. Algorithm 2.4.1 selects two principal vertices generating those extreme points.

2.7. Hurwitz Stability of Weighted Diamonds

The principal vertices technique that has been developed so far for box polynomials is applicable to general polytopes of polynomials with real or complex coefficients. In this section we apply the principal vertices technique to diamond polynomials with real coefficients. Diamond polynomial families (sometimes referred to as dual to box polynomial families) have been investigated by Tempo [Te], Bose and Kim [BoK], Barmish, Tempo, Hollot and Kang [BTHK], Foo and Soh [FS2], Katbab and Jury [KJ1], and Wang and Huang [WH] for the special case when all the exposed edges of the diamond are of equal size. Different types of "edge", "extreme point", and mixed "edge–extreme point" criteria have been developed. In this section we consider diamond families without the equal size restriction. We develop edge–type stability conditions, and in particular identify the minimal set of exposed edges of the diamond whose stability guarantees the stability of the entire polynomial family. The identification leads to extreme point results presented in the next section.

Consider the polynomial family

$$\mathbf{P} = \{\mathbf{p}(s, \mathbf{x}) \ : \ \mathbf{p}(s, \mathbf{x}) = a_0(\mathbf{x}) + a_1(\mathbf{x})s + \ldots + a_{n-1}(\mathbf{x})s^{n-1} + s^n, \ \mathbf{x} \in \mathbf{D}\}, \quad (2.7.1)$$

where $a_i(x) = \underline{a}_i + \alpha_i x_{i+1}$, $\alpha_i > 0$, $i = 0, \ldots, n-1$, $n \geq 2$, and

$$\mathbf{D} = \left\{\mathbf{x} \ : \ \mathbf{x} \in \mathbf{R}^n, \ \sum_{i=1}^{n} |x_i| \leq 1\right\}.$$

The $2n$ vertices of the diamond $\mathbf{D}$ are denoted by

$$\mathbf{V} = \left\{\pm\mathbf{v}^1, \pm\mathbf{v}^2, \ldots, \pm\mathbf{v}^n\right\}, \text{ where } \left\{\mathbf{v}^1, \ldots, \mathbf{v}^n\right\} \text{ is the standard basis for } \mathbf{R}^n.$$

For the vertices $\mathbf{v}^i$ and $-\mathbf{v}^i$ the sets of the $2(n-1)$ neighboring vertices are

$$\mathbf{N}_{\mathbf{v}^i} = \mathbf{N}_{-\mathbf{v}^i} = \left\{\pm\mathbf{v}^k\right\}_{k \neq i}.$$

For a fixed real w consider the affine mapping from $\mathbf{R}^n$ to the complex plane $\mathbf{C}$ defined by $\mathbf{x} \to f(\mathbf{x})$, where

$$f(\mathbf{x}) = \mathbf{p}(jw, \mathbf{x}) = [\underline{a}_0 + \alpha_0 x_1] + \ldots + [\underline{a}_{n-1} + \alpha_{n-1} x_n](jw)^{n-1} + (jw)^n. \tag{2.7.2}$$

The partial derivatives of f are $\alpha_{k-1}(jw)^{k-1}$, $k = 1, \ldots n$. The n partial derivatives generate *at most* four different directions.

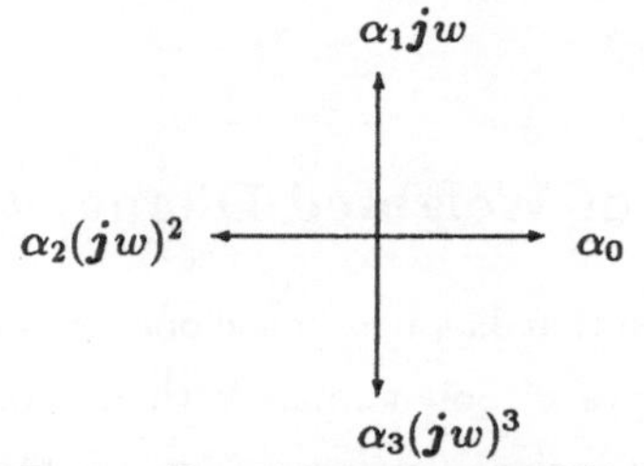

Four different directions generated by the partial derivatives

The vertex $\mathbf{v}^i$ ($-\mathbf{v}^i$ respectively) is a principal vertex of the diamond if and only if the half lines generated by at most $2(n-1)$ nonzero complex numbers

$$\left\{\pm\frac{\partial f}{\partial x_k} - \frac{\partial f}{\partial x_i}\right\}_{k \neq i} \qquad \left(\left\{\pm\frac{\partial f}{\partial x_k} + \frac{\partial f}{\partial x_i}\right\}_{k \neq i} \text{ respectively}\right) \tag{2.7.3}$$

are contained in an open half plane. Relations (2.7.3) clearly indicate that $\mathbf{v}^i$ is a principal vertex if and only if $-\mathbf{v}^i$ is a principal vertex.

The vertex $\mathbf{v}^1$ is a principal vertex of the diamond $\mathbf{D}$ if and only if the nonzero complex numbers

$$\left\{\pm\alpha_k(jw)^k - \alpha_0\right\}_{k \neq 0} \tag{2.7.4}$$

are contained in an open half plane. If $w = 0$ the set (2.7.4) contains a single number $-\alpha_0$, hence in this case $\mathbf{v}^1$ is a principal vertex. When $w > 0$ the complex numbers

$$\left\{\pm\alpha_k(jw)^k - \alpha_0\right\}_{\text{odd } k} \tag{2.7.5}$$

are located in the left–half plane.

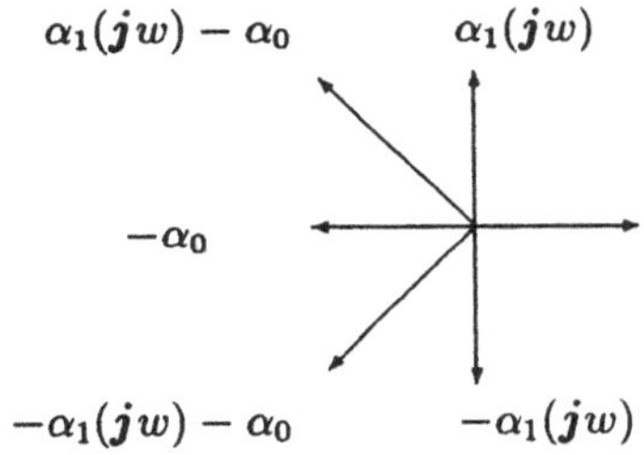

$\left\{\pm\alpha_k(jw)^k - \alpha_0\right\}_{\text{odd } k}$

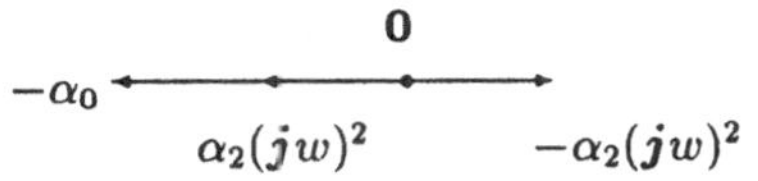

$\left\{\pm\alpha_k(jw)^k - \alpha_0\right\}_{\text{even } k>0}$

The complex numbers

$$\left\{\pm\alpha_k(jw)^k - \alpha_0\right\}_{\text{even } k>0} \tag{2.7.6}$$

are located in the left–half plane if and only if

$$\alpha_{2k}w^{2k} \leq \alpha_0 \qquad \text{for each} \qquad k = 1, 2, \ldots, m, \tag{2.7.7}$$

where $2m$ is the maximal integer not exceeding $n-1$, i.e., $2m \leq n-1 < 2m+2$. Let

$$\underline{w}_0 = 0, \text{ and } \overline{w}_0 = \min\left\{\left(\frac{\alpha_0}{\alpha_2}\right)^{\frac{1}{2}}, \left(\frac{\alpha_0}{\alpha_4}\right)^{\frac{1}{4}}, \ldots, \left(\frac{\alpha_0}{\alpha_{2m}}\right)^{\frac{1}{2m}}\right\}, \tag{2.7.8}$$

then $\pm\mathbf{v}^1$ are principal vertices if and only if $w \in [\underline{w}_0, \overline{w}_0]$.

Similar arguments show that $\pm\mathbf{v}^2$ are principal vertices if and only if $w \in (\underline{w}_1, \overline{w}_1]$, where

$$\underline{w}_1 = 0, \text{ and } \overline{w}_1 = \min\left\{\left(\frac{\alpha_1}{\alpha_3}\right)^{\frac{1}{2}}, \left(\frac{\alpha_1}{\alpha_5}\right)^{\frac{1}{4}}, \ldots, \left(\frac{\alpha_1}{\alpha_{2l+1}}\right)^{\frac{1}{2l}}\right\}, \tag{2.7.9}$$

and $2l+1$ is the maximal odd integer not exceeding $n-1$. Furthermore, the vertices $\pm\mathbf{v}^{2i+1}$, $i = 1, 2, \ldots, m$ are principal vertices if and only if the following conditions hold:

1. $\underline{w}_{2i} < \overline{w}_{2i}$,
2. $w \in [\underline{w}_{2i}, \overline{w}_{2i}]$,

where

$$\begin{aligned} \underline{w}_{2i} &= \max\left\{\left(\tfrac{\alpha_0}{\alpha_{2i}}\right)^{\frac{1}{2i}}, \ldots, \left(\tfrac{\alpha_{2(i-1)}}{\alpha_{2i}}\right)^{\frac{1}{2}}\right\} \\ \overline{w}_{2i} &= \min\left\{\left(\tfrac{\alpha_{2i}}{\alpha_{2(i+1)}}\right)^{\frac{1}{2}}, \ldots, \left(\tfrac{\alpha_{2i}}{\alpha_{2m}}\right)^{\frac{1}{2m-2i}}\right\}. \end{aligned} \tag{2.7.10}$$

Analogously, the vertices $\pm \mathbf{v}^{2i+2}$, $i = 1, 2, \ldots, l$ are principal vertices if and only if the following conditions hold:

1. $\underline{w}_{2i+1} < \overline{w}_{2i+1}$,

2. $w \in [\underline{w}_{2i+1}, \overline{w}_{2i+1}]$,

where

$$\begin{aligned} \underline{w}_{2i+1} &= \max\left\{\left(\frac{\alpha_1}{\alpha_{2i+1}}\right)^{\frac{1}{2i}}, \ldots, \left(\frac{\alpha_{2i-1}}{\alpha_{2i+1}}\right)^{\frac{1}{2}}\right\} \\ \overline{w}_{2i+1} &= \min\left\{\left(\frac{\alpha_{2i+1}}{\alpha_{2(i+1)+1}}\right)^{\frac{1}{2}}, \ldots, \left(\frac{\alpha_{2i+1}}{\alpha_{2l+1}}\right)^{\frac{1}{2l-2i}}\right\}. \end{aligned} \tag{2.7.11}$$

Let $E \subseteq \{0, 2, \ldots, 2m\}$ be the set of even indices so that $\underline{w}_e < \overline{w}_e$. Analogously define the set $O \subseteq \{1, 3, \ldots, 2l+1\}$ so that $\underline{w}_o < \overline{w}_o$. Then

$$\bigcup_{e \in E} [\underline{w}_e, \overline{w}_e] = \mathbf{R}_+, \text{ and if } e_1 \neq e_2, \text{ then } (\underline{w}_{e_1}, \overline{w}_{e_1}) \bigcap (\underline{w}_{e_2}, \overline{w}_{e_2}) = \emptyset. \tag{2.7.12}$$

Analogously

$$\bigcup_{o \in O} [\underline{w}_o, \overline{w}_o] = \mathbf{R}_+, \text{ and if } o_1 \neq o_2, \text{ then } (\underline{w}_{o_1}, \overline{w}_{o_1}) \bigcap (\underline{w}_{o_2}, \overline{w}_{o_2}) = \emptyset. \tag{2.7.13}$$

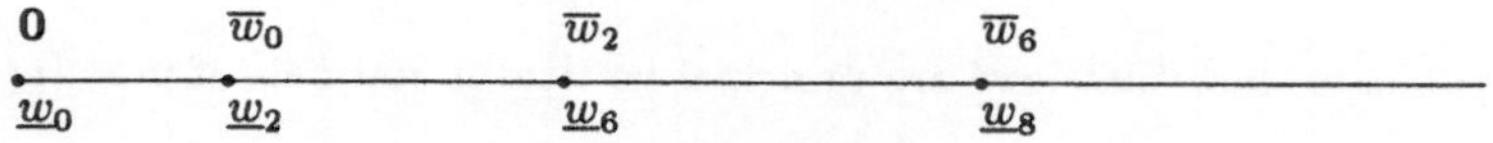

Even partition of $\mathbf{R}_+$

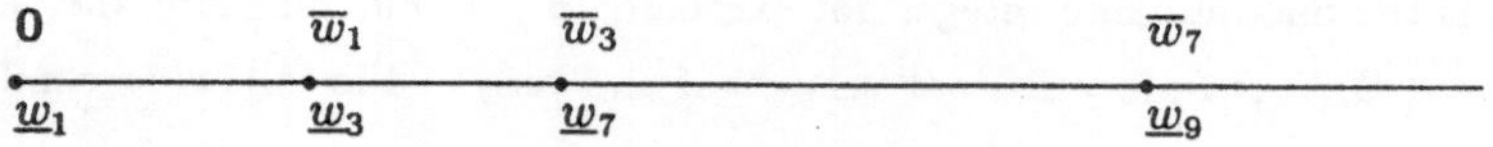

Odd partition of $\mathbf{R}_+$

For each $w > 0$ there exist indices $e \in E$, and $o \in O$ so that $w \in [\underline{w}_e, \overline{w}_e] \bigcap [\underline{w}_o, \overline{w}_o]$, and the four vertices $\{\pm \mathbf{v}^{e+1}, \pm \mathbf{v}^{o+1}\}$ constitute the set of principal vertices at w. When $w = 0$ the set of principal vertices is $\left\{\pm \mathbf{v}^1\right\}$. Let $\mathbf{p}_c(s) = \underline{a}_0 + \underline{a}_1 s + \ldots + \underline{a}_{n-1} s^{n-1} + s^n$. For each $w \geq 0$ the value set $\mathcal{P}_w$ is a parallelogram with the diagonals parallel to the coordinate axes.

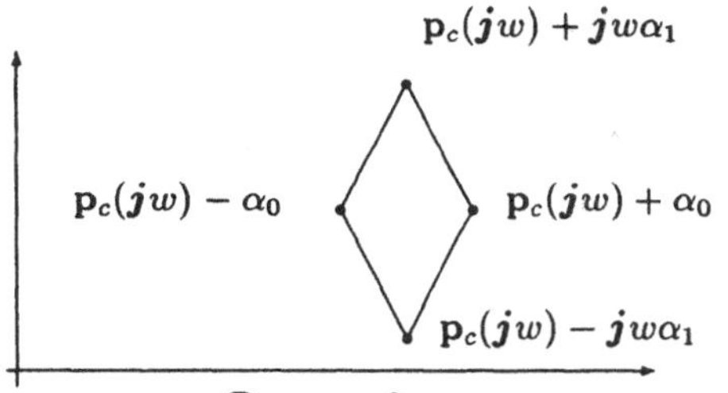

$\mathcal{P}_w$, $w > 0$

the principal vertices are

$\pm\mathbf{v}^1$, $\pm\mathbf{v}^2$

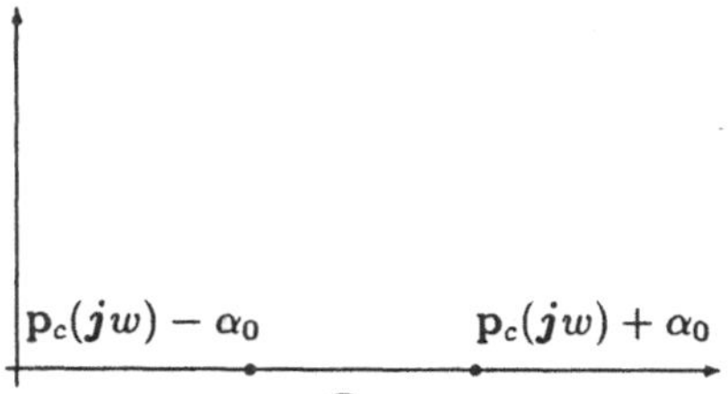

$\mathcal{P}_0$

the principal vertices are

$\pm\mathbf{v}^1$

For $w > 0$ we choose $e \in E$, and $o \in O$ so that $w \in [\underline{w}_e, \overline{w}_e] \bigcap [\underline{w}_o, \overline{w}_o]$. The value set $\mathcal{P}_w$ is generated by the four complex numbers $\left\{\mathbf{p}\left(jw, \pm\mathbf{v}^{e+1}\right), \mathbf{p}\left(jw, \pm\mathbf{v}^{o+1}\right)\right\}$, and the stability of the four edges

$$\pm\, \lambda\mathbf{v}^{e+1} \pm (1-\lambda)\mathbf{v}^{o+1}, \qquad 0 \le \lambda \le 1 \tag{2.7.14}$$

guarantees the stability of the diamond family (2.7.1) on $[\underline{w}_e, \overline{w}_e] \bigcap [\underline{w}_o, \overline{w}_o]$. At $w = 0$ the stability of the family is guaranteed if $a_0 - \alpha_1 > 0$. Let the numbers

$$\bigcup_{e \in E,\, o \in O} \{\underline{w}_e, \overline{w}_e, \underline{w}_o, \overline{w}_o\} \tag{2.7.15}$$

be written in the increasing order

$$0 = w_0 < w_1 < \ldots < w_M < w_{M+1} = \infty. \tag{2.7.16}$$

We shall show that the diamond polynomial family (2.7.1) is stable if and only if *at most* $4(M+1)$ one dimensional edges of the diamond $\mathbf{D}$ are Hurwitz stable. The sets E and O determine the set of the edges whose stability implies the stability of the entire polynomial family. In order to clarify the concepts introduced so far we provide two illustrative examples.

Example 2.7.1 Consider the diamond polynomial family (2.7.1) of degree 5 with the weights

$$\alpha_0 = 1,\ \alpha_1 = 2,\ \alpha_2 = 3,\ \alpha_3 = 4,\ \alpha_4 = 5.$$

Then

$$\underline{w}_0 = 0,\ \overline{w}_0 = \left(\frac{1}{3}\right)^{\frac{1}{2}},\ \underline{w}_2 = \left(\frac{1}{3}\right)^{\frac{1}{2}},\ \overline{w}_2 = \left(\frac{3}{5}\right)^{\frac{1}{2}},\ \underline{w}_4 = \left(\frac{3}{5}\right)^{\frac{1}{2}},\ \overline{w}_4 = \infty,$$

and

$$\underline{w}_1 = 0,\ \overline{w}_1 = \left(\frac{1}{2}\right)^{\frac{1}{2}},\ \underline{w}_3 = \left(\frac{1}{2}\right)^{\frac{1}{2}},\ \overline{w}_3 = \infty.$$

$E = \{0, 2, 4\}$, $O = \{1, 3\}$, the partition (2.7.16) is

$$w_0 = 0 < w_1 = \left(\frac{1}{3}\right)^{\frac{1}{2}} < w_2 = \left(\frac{1}{2}\right)^{\frac{1}{2}} < w_3 = \left(\frac{3}{5}\right)^{\frac{1}{2}} < w_4 = \infty, \text{ and } M = 3.$$

Example 2.7.2 Consider the diamond polynomial family (2.7.1) of degree 6 with the weights

$$\alpha_0 = 9,\ \alpha_1 = 4,\ \alpha_2 = 9,\ \alpha_3 = 1,\ \alpha_4 = 1,\ \alpha_5 = \frac{1}{4}.$$

In this case

$$\underline{w}_0 = 0,\ \overline{w}_0 = 1,\ \underline{w}_2 = 1,\ \overline{w}_2 = 3,\ \underline{w}_4 = 3,\ \overline{w}_4 = \infty,$$

and

$$\underline{w}_1 = 0,\ \overline{w}_1 = 2,\ \underline{w}_3 = 2,\ \overline{w}_3 = 2,\ \underline{w}_5 = 2,\ \overline{w}_5 = \infty.$$

Hence $E = \{0, 2, 4\}$, $O = \{1, 5\}$, and the partition (2.7.16) is:

$$w_0 = 0 < w_1 = 1 < w_2 = 2 < w_5 = 3 < w_4 = \infty, \text{ and } M = 3.$$

Theorem 2.7.1 The polynomial family (2.7.1) is Hurwitz stable if and only if at most $4(M+1)$ one dimensional edges of the diamond $\mathbf{D}$

$$\left\{\pm\lambda \mathbf{p}(s, \mathbf{v}^{e+1}) \pm (1-\lambda)\mathbf{p}(s, \mathbf{v}^{o+1})\right\}, \quad e \in E,\ o \in O,\ (\underline{w}_e, \overline{w}_e) \bigcap (\underline{w}_o, \overline{w}_o) \neq \emptyset \tag{2.7.17}$$

are Hurwitz stable.

Proof: Suppose that there exists an unstable polynomial $\mathbf{p}(s) \in \mathbf{P}$. According to the "zero exclusion criterion" there exists $w' \geq 0$, and $\mathbf{p}'(s) \in \mathbf{P}$ so that $\mathbf{p}'(jw') = 0$, and $0 \in \mathcal{P}_{w'}$. The condition $0 \notin \mathcal{P}_0$ leads to the existence of $0 < w < w'$ so that $0 \in \partial\mathcal{P}_w$. Let $e \in E$, and $o \in O$ be such that

$$w \in [\underline{w}_e, \overline{w}_e] \bigcap [\underline{w}_o, \overline{w}_o].$$

The boundary of the value set $\partial\mathcal{P}_w$ is maid up of four segments

$$\pm\lambda \mathbf{p}(jw, \mathbf{v}^{e+1}) \pm (1-\lambda)\mathbf{p}(jw, \mathbf{v}^{o+1}), \qquad 0 \leq \lambda \leq 1.$$

Due to the assumptions of the theorem the segments do not contain the origin. This contradiction completes the proof.

2.8. Extreme Point Results

In this section we consider a convex stability region Ω. Let $\delta : I_\Omega \to \partial\Omega$ be the parameterization in terms of arc length of the boundary of the stability region Ω so that the principal normal is pointing inside Ω. That is at each $w \in I_\Omega$ where $\delta(w)$ is differentiable one has

$$|\delta'(w)| = 1, \quad \text{and} \quad \delta''(w) = \boldsymbol{j}\kappa(w)\delta'(w) \text{ with } \kappa(w) > 0.$$

For example, in the sector stability case which is the main subject of the section, $I_\Omega = (-\infty, \infty)$, and

$$\delta(w) = \begin{cases} we^{\boldsymbol{j}\phi} & \text{if } w \geq 0 \\ we^{\boldsymbol{j}(2\pi-\phi)} & \text{if } w < 0 \end{cases}.$$

Extreme point results derived in this section are concerned with box and diamond polynomial families. First, the extreme point results discovered by Foo and Soh [FS1] are recovered, and, in particular, Kharitonov's theorem is obtained as a special case. Next, extreme point results for weighted diamond polynomials are provided.

We start with a definition and two auxiliary results.

Definition 2.8.1 *The dot product of two complex numbers $z_1 = x_1 + \boldsymbol{j}y_1$, and $z_2 = x_2 + \boldsymbol{j}y_2$ is*

$$\langle z_1, z_2 \rangle = x_1x_2 + y_1y_2.$$

We remark that the following simple relation holds for each $w \in I_\Omega$, and s in the interior of the region Ω

$$\langle -\boldsymbol{j}\delta'(w), \delta(w) - s \rangle > 0. \tag{2.8.1}$$

Lemma 2.8.1 *Let $\mathbf{p}(s) = (s - s_1) \times \ldots \times (s - s_n)$ be Ω stable polynomial. Then*

$$\forall w \in I_\Omega^+ \text{ one has } \frac{d}{dw} \arg \mathbf{p}(\delta(w)) > 0.$$

Proof:

$$\begin{aligned} \frac{d}{dw}\arg \mathbf{p}(\delta(w)) &= \frac{d}{dw}\text{Im } \log \mathbf{p}(\delta(w)) = \text{Im } \delta'(w)\frac{\mathbf{p}'(\delta(w))}{\mathbf{p}(\delta(w))} \\ &= \text{Im} \sum_{k=1}^{n} \frac{\delta'(w)}{\delta(w) - s_k} = \sum_{k=1}^{n} \text{Im} \frac{\delta'(w)\overline{(\delta(w) - s_k)}}{|\delta(w) - s_k|^2} \\ &= \sum_{k=1}^{n} \frac{\langle -\boldsymbol{j}\delta'(w), \delta(w) - s \rangle}{|\delta(w) - s_k|^2} > 0. \end{aligned}$$

Lemma 2.8.2 *(see [Ra2], Lemma 10.) If $\mu \mathbf{p}_1(\delta(w)) + (1-\mu)\mathbf{p}_0(\delta(w)) = 0$, then*

$$\frac{d}{dw} arg\ [\mathbf{p}_1(\delta(w)) - \mathbf{p}_0(\delta(w))] = \mu \frac{d}{dw} arg\ \mathbf{p}_0(\delta(w)) + (1-\mu)\frac{d}{dw} arg\ \mathbf{p}_1(\delta(w)).$$

Proof:

$$\begin{aligned}
\frac{d}{dw}\text{arg } [\mathbf{p}_1(\delta(w)) - \mathbf{p}_0(\delta(w))] &= \frac{d}{dw}\text{Im } \log[\mathbf{p}_1(\delta(w)) - \mathbf{p}_0(\delta(w))] \\
&= \text{Im } \delta'(w)\frac{\mathbf{p}_1'(\delta(w)) - \mathbf{p}_0'(\delta(w))}{\mathbf{p}_1(\delta(w)) - \mathbf{p}_0(\delta(w))} \\
&= \text{Im } \delta'(w)\left[\frac{\mathbf{p}_1'(\delta(w))}{\mathbf{p}_1(\delta(w)) - \mathbf{p}_0(\delta(w))} - \frac{\mathbf{p}_0'(\delta(w))}{\mathbf{p}_1(\delta(w)) - \mathbf{p}_0(\delta(w))}\right] \\
&= \mu\text{Im } \delta'(w)\frac{\mathbf{p}_0'(\delta(w))}{\mathbf{p}_0(\delta(w))} + (1-\mu)\text{Im } \delta'(w)\frac{\mathbf{p}_1'(\delta(w))}{\mathbf{p}_1(\delta(w))} \\
&= \mu\frac{d}{dw}\text{arg } \mathbf{p}_0(\delta(w)) + (1-\mu)\frac{d}{dw}\text{arg } \mathbf{p}_1(\delta(w)).
\end{aligned}$$

Lemma 2.8.2 is the main technical tool that leads to extreme point results provided in the remainder of the chapter.

2.8.1. Sector Stability of Box Polynomials

Consider the polynomial family

$$\mathbf{P} = \left\{\mathbf{p}(s, \mathbf{x}) = [\underline{a}_0 + \alpha_0 x_1] + \ldots + [\underline{a}_{n-1} + \alpha_{n-1} x_n]s^{n-1} + s^n,\ \mathbf{x} \in \mathbf{B}\right\}, \qquad (2.8.2)$$

and the sector stability region

$$\Omega = \left\{s\ :\ s = re^{\mathbf{j}\theta},\ |\theta - \pi| \leq \phi_0,\ r \geq 0\right\}, \qquad \phi_0 \leq \frac{\pi}{2}. \qquad (2.8.3)$$

The set of principal directions at $w > 0$ is generated by the complex numbers

$$\pm\,\alpha_0,\ \pm\alpha_1 w e^{\mathbf{j}\phi},\ \pm\alpha_2 w^2 e^{2\mathbf{j}\phi},\ \ldots, \pm\alpha_{n-1} w^{n-1} e^{(n-1)\mathbf{j}\phi},\ \phi = \pi - \phi_0. \qquad (2.8.4)$$

The set of the principal directions does not depend on w, and the number of the principal directions does not exceed $2n$. In other words $n(w) = k$ for each $w > 0$, $n \geq k$, and the set of principal vertices

$$\left\{\mathbf{v}^1, \ldots, \mathbf{v}^{2k}\right\} \qquad (2.8.5)$$

is frequency independent. Furthermore the arguments of the complex numbers $D_i(w)$ do not depend on w for each $i = 1, \ldots, 2k$. This observation makes application of (2.4.8) possible. We now ready to state and prove the extreme point result.

Theorem 2.8.1 The polynomial family (2.8.2) is sector stable if and only if the $2k$ principal vertices (2.8.5) are sector stable.

Proof: Suppose there exists an unstable $\mathbf{p}(s) \in \mathbf{P}$. Since $\mathbf{p}(s, \mathbf{v}^1)$ is stable $\underline{a}_0 > 0$, and $0 \notin \mathcal{P}_0$. Hence there exists $w > 0$ so that $0 \in \partial\mathcal{P}_w$. Let

$$\left\{\lambda \mathbf{p}(\delta(w), \mathbf{v}^{i+1}) + (1-\lambda)\mathbf{p}(\delta(w), \mathbf{v}^i), 0 \le \lambda \le 1\right\}$$

be the edge of $\mathcal{P}_w$ that hits the origin. Since $\mathbf{p}(s, \mathbf{v}^{i+1})$, and $\mathbf{p}(s, \mathbf{v}^i)$ are stable polynomials there exists $0 < \mu < 1$ so that

$$\mu \mathbf{p}(\delta(w), \mathbf{v}^{i+1}) + (1-\mu)\mathbf{p}(\delta(w), \mathbf{v}^i) = 0.$$

According to Lemma 2.8.2

$$\begin{aligned} \frac{d}{dw}\arg\left[\mathbf{p}(\delta(w), \mathbf{v}^{i+1}) - \mathbf{p}(\delta(w), \mathbf{v}^i)\right] &= \mu\frac{d}{dw}\arg \mathbf{p}(\delta(w), \mathbf{v}^i) \\ &+ (1-\mu)\frac{d}{dw}\arg \mathbf{p}(\delta(w), \mathbf{v}^{i+1}). \end{aligned}$$

Due to (2.4.8) the LHS of this expression is zero, on the other hand, due to Lemma 2.8.1 the RHS of the expression is positive. This contradiction completes the proof.

Remark 2.8.1 When $\phi_0 = \frac{\pi}{2}$ the number of principal vertices is 4 (see Example 2.6.2), and Kharitonov's Theorem is a particular case of Theorem 2.8.1.

2.8.2. Hurwitz Stability of Diamond Polynomials

In this subsection we present extreme point results for polynomial families

$$\mathbf{P} = \left\{\mathbf{p}(s, \mathbf{x}) = [\underline{a}_0 + \alpha_0 x_1] + \ldots + [\underline{a}_{n-1} + \alpha_{n-1} x_n]s^{n-1} + s^n,\ \mathbf{x} \in \mathbf{D}\right\}, \qquad (2.8.6)$$

and the stability region Ω = the left half plane. The parameterization of the boundary is $\delta(w) = \boldsymbol{j}w$.

In contrast with Theorem 2.2.1 extreme point results do *not* hold for arbitrary weighted diamond polynomials. First extreme point results for weighted diamond polynomials are due to Kharitonov and Tempo [KT]. In this section we provide an "interlacing" condition under which the extreme point results hold. Our focus is slightly different from that of [KT].

Each $w \ge 0$ is covered by two intervals $[\underline{w}_e, \overline{w}_e]$, and $[\underline{w}_o, \overline{w}_o]$. The main result of this subsection shows that the stability of the vertices

$$\left\{\pm\mathbf{v}^{e+1},\ \pm\mathbf{v}^{o+1}\right\},\ e \in E,\ o \in O \qquad (2.8.7)$$

implies the stability of the diamond polynomial family when the indices e and o are not far apart.

Theorem 2.8.2 Suppose that for each $w > 0$ there exist "covering" intervals $[\underline{w}_e, \overline{w}_e]$, and $[\underline{w}_o, \overline{w}_o]$ such that

1. $w \in [\underline{w}_e, \overline{w}_e] \bigcap [\underline{w}_o, \overline{w}_o]$.

2. $|e - o| = 1$.

Then the Hurwitz stability of $2\,|E + O|$ principal vertices (2.8.7) implies stability of the diamond polynomial family $\mathbf{P}$.

The proof of the theorem is based on a number of auxiliary results given next.

Lemma 2.8.3 *(see [J2], p. 61.) Let*

$$\mathbf{p}(s) = a_0 + a_1 s + \ldots + a_{n-1}s^{n-1} + s^n = u(s^2) + sv(s^2)$$

be a polynomial with real coefficients, where

$$u(s^2) = a_0 + a_2 s^2 + \ldots, \; and \; v(s^2) = a_1 + a_3 s^2 + \ldots.$$

The following three conditions are equivalent:

1. $\mathbf{p}(s)$ *is Hurwitz stable.*

2. $u(\boldsymbol{j}s) + \boldsymbol{j}v(\boldsymbol{j}s)$ *is Hurwitz stable.*

3. $u(-\boldsymbol{j}s) + sv(-\boldsymbol{j}s)$ *is Hurwitz stable.*

The next lemma is a particular case of Lemma 2, [BHKT] .

Lemma 2.8.4 *Let* $\mathbf{p}(s) = a_0 + a_1 s + \ldots + a_{n-1}s^{n-1} + s^n$ *be a polynomial with real coefficients,* α', α'' *are real numbers, and* $k \in \{0, \ldots, n-1\}$. *The polynomial segment*

$$\mathbf{p}_\lambda(s) = \left\{\mathbf{p}(s) + \lambda\left[\alpha' s^k + \alpha'' s^{k+1}\right]\right\}, \qquad 0 \le \lambda \le 1$$

is Hurwitz stable if and only if the extreme polynomials $\mathbf{p}_0(s)$ *and* $\mathbf{p}_1(s)$ *are Hurwitz stable.*

Proof: We write $\mathbf{p}_\lambda(s) = u_\lambda(s^2) + sv_\lambda(s^2)$ and consider the cases of even and odd k.
Case 1. $k = 2i$.
In this case

$$u_\lambda(s^2) = u(s^2) + \lambda\alpha' s^{2i}, \text{ and } v_\lambda(s^2) = v(s^2) + \lambda\alpha'' s^{2i}.$$

According to Lemma 2.8.3 the polynomials $\mathbf{p}_\lambda(s)$ are stable if and only if the polynomials $\overline{\mathbf{p}}_\lambda(s) = u_\lambda(\boldsymbol{j}s) + \boldsymbol{j}v_\lambda(\boldsymbol{j}s)$ are stable. The stability of $\mathbf{p}_\lambda(s)$, $\lambda = 0, 1$ implies stability of the end polynomials $\overline{\mathbf{p}}_\lambda(s)$, $\lambda = 0, 1$. If there exists λ^* so that $\overline{\mathbf{p}}_{\lambda^*}(s)$ is unstable, then there exists $\mu \in (0,1)$, and $w > 0$ so that

$$0 = \overline{\mathbf{p}}_\mu(\boldsymbol{j}w) = \mu\overline{\mathbf{p}}_1(\boldsymbol{j}w) + (1-\mu)\overline{\mathbf{p}}_0(\boldsymbol{j}w).$$

We now invoke Lemma 2.8.2 with $\overline{\mathbf{p}}_1(s)$, $\overline{\mathbf{p}}_0(s)$, $\overline{\mathbf{p}}_1(\boldsymbol{j}w) - \overline{\mathbf{p}}_0(\boldsymbol{j}w) = (\mu\alpha' + \boldsymbol{j}\mu\alpha'')(\boldsymbol{j}w)^i$, and obtain

$$\begin{aligned} 0 &= \frac{d}{dw}\arg\left[(\boldsymbol{j}w)^i\right] \\ &= \frac{d}{dw}\arg\left[(\mu\alpha' + \boldsymbol{j}\mu\alpha'')(\boldsymbol{j}w)^i\right] \\ &= (1-\mu)\frac{d}{dw}\arg\overline{\mathbf{p}}_1(\boldsymbol{j}w) + \mu\frac{d}{dw}\arg\overline{\mathbf{p}}_0(\boldsymbol{j}w). \end{aligned}$$

According to Lemma 2.8.1 the right hand side of the expression is positive. This contradiction completes the proof of case 1.

Case 2. $k = 2i + 1$.
In this case

$$u_\lambda(s^2) = u(s^2) + \lambda\alpha'' s^{2i+2}, \text{ and } v_\lambda(s^2) = v(s^2) + \lambda\alpha' s^{2i}.$$

According to Lemma 2.8.3 the polynomials $\mathbf{p}_\lambda(s)$ are stable if and only if the polynomials $\overline{\mathbf{p}}_\lambda(s) = u_\lambda(-\boldsymbol{j}s) + sv_\lambda(-\boldsymbol{j}s)$ are stable. The rest of the proof repeats the proof of the case 1 with

$$\overline{\mathbf{p}}_1(\boldsymbol{j}w) - \overline{\mathbf{p}}_0(\boldsymbol{j}w) = \left[\mu\alpha'(-\boldsymbol{j})^i + \mu\alpha''(-\boldsymbol{j})^{i+1}\right](\boldsymbol{j}w)^{i+1},$$

when

$$0 = \mu\overline{\mathbf{p}}_1(\boldsymbol{j}w) + (1-\mu)\overline{\mathbf{p}}_0(\boldsymbol{j}w).$$

To complete the proof of Theorem 2.8.2 we note that

$$\pm\mathbf{p}(s, \mathbf{v}^{o+1}) \mp \mathbf{p}(s, \mathbf{v}^{e+1}) = \pm\alpha_o s^o \mp \alpha_e s^e,$$

and the assumption $|e - o| = 1$ makes application of Lemma 2.8.4 possible. This completes the proof of the theorem.

Finally we provide an example of diamond families whose stability is guaranteed by stability of eight vertex polynomials only.

Example 2.8.1 Let γ, and α be positive real numbers. Consider a diamond polynomial family

$$\left\{[\underline{a}_0+\alpha_0x_1]+[\underline{a}_1+\alpha_1x_2]s+[\underline{a}_2+\alpha_2x_3]s^2+\ldots+[\underline{a}_{n-1}+\alpha_{n-1}x_n]s^{n-1}+s^n,\ \mathbf{x}\in\mathbf{D}\right\},$$

where $\alpha_k=\gamma\alpha^k$. Then

$$\underline{w}_0=0,\ \overline{w}_0=\frac{1}{\alpha}=\underline{w}_2=\overline{w}_2=\ldots=\underline{w}_{2(m-1)}=\overline{w}_{2(m-1)}=\underline{w}_{2m}=\frac{1}{\alpha},\ \text{ and } \overline{w}_{2m}=\infty.$$

Analogously

$$\underline{w}_1=0,\ \overline{w}_1=\frac{1}{\alpha},\ \underline{w}_3=\overline{w}_3=\underline{w}_{2l-1}=\overline{w}_{2l-1}=\underline{w}_{2l+1}=\frac{1}{\alpha},\ \text{ and } \overline{w}_{2l+1}=\infty.$$

The index sets $E=\{0,2m\}$, and $O=\{1,2l+1\}$. Conditions of Theorem 2.8.2 are met, and Hurwitz stability of the eight vertices

$$\pm\mathbf{v}^1,\ \pm\mathbf{v}^2,\ \pm\mathbf{v}^{2m+1},\ \pm\mathbf{v}^{2l+1}$$

implies stability of the entire polynomial family.

Remark 2.8.2 While the polynomial family of Example 2.7.2 fails to meet the assumptions of Theorem 2.8.2, the polynomial family of Example 2.7.1 satisfies the assumptions. Hence stability of 10 vertex polynomials guarantees the stability of the polynomial family in Example 2.7.1. Example 2.7.1 is a particular case of the diamond polynomial family

$$\left\{[\underline{a}_0+\alpha_0x_1]+[\underline{a}_1+\alpha_1x_2]s+[\underline{a}_2+\alpha_2x_3]s^2+\ldots+[\underline{a}_{n-1}+\alpha_{n-1}x_n]s^{n-1}+s^n,\ \mathbf{x}\in\mathbf{D}\right\},$$

where $\alpha_k=\gamma(k+1)$. Elementary calculus shows that the function $\psi_y(x)=\left(\frac{x}{y}\right)^{\frac{1}{y-x}}$ satisfies

$$\psi_y(x_1)<\psi_y(x_2) \text{ when } 0<x_1<x_2,$$
$$\text{and } \frac{k}{k+2}<\frac{k+1}{k+3},\ k=1,2,\ldots \tag{2.8.8}$$

A straightforward application of (2.8.8) to (2.7.10) and (2.7.11) with $\alpha_k=\gamma(k+1)$ yields

$$\underline{w}_{2i}=\left(\frac{2i-1}{2i+1}\right)^{\frac{1}{2}}<\underline{w}_{2i+1}=\left(\frac{2i}{2i+2}\right)^{\frac{1}{2}}<\overline{w}_{2i}=\left(\frac{2i+1}{2i+3}\right)^{\frac{1}{2}}$$
$$<\overline{w}_{2i+1}=\left(\frac{2i+2}{2i+4}\right)^{\frac{1}{2}}. \tag{2.8.9}$$

Relations (2.8.9) show that the assumptions of Theorem 2.8.2 are satisfied, and the polynomial family is Hurwitz stable if and only if *all* $2n$ vertices of the diamond are Hurwitz stable.

Finally we show by an example that the extreme point result may fail if conditions of Theorem 2.8.2 are violated.

Example 2.8.2 (see also [KT]) Consider the diamond polynomial family of degree 5

$$\left\{ \begin{array}{l} [1000 + 500x_1] + [1250 + 50x_2]\, s + [1350 + 50x_3]\, s^2 + \\ [1000 + 500x_4]\, s^3 + [100 + 10x_5]\, s^4 + s^5 \end{array} \right\}.$$

A straightforward evaluation of the roots using MATLAB shows that all the vertex polynomials are Hurwitz, but the polynomial

$$\mathbf{p}(s, \mathbf{x}^*) = 1300 + 1250s + 1350s^2 + 1200s^3 + 100s^4 + s^5, \ \mathbf{x}^* = (0.6, 0, 0, 0.4, 0)^t$$

is *not* Hurwitz. A direct application of (2.7.10) and (2.7.11) yields the following relations:

$$\underline{w}_0 = 0, \ \overline{w}_0 = (50)^{\frac{1}{4}}, \ \underline{w}_2 = (10)^{\frac{1}{2}}, \ \overline{w}_2 = (5)^{\frac{1}{2}}, \ \underline{w}_4 = (50)^{\frac{1}{4}}, \ \overline{w}_4 = \infty,$$

and

$$\underline{w}_1 = 0, \ \overline{w}_1 = \left(\frac{1}{10}\right)^{\frac{1}{2}}, \ \underline{w}_3 = \left(\frac{1}{10}\right)^{\frac{1}{2}}, \ \overline{w}_3 = \infty.$$

For this example

$$E = \{0, 4\}, \ \text{ and } O = \{1, 3\}.$$

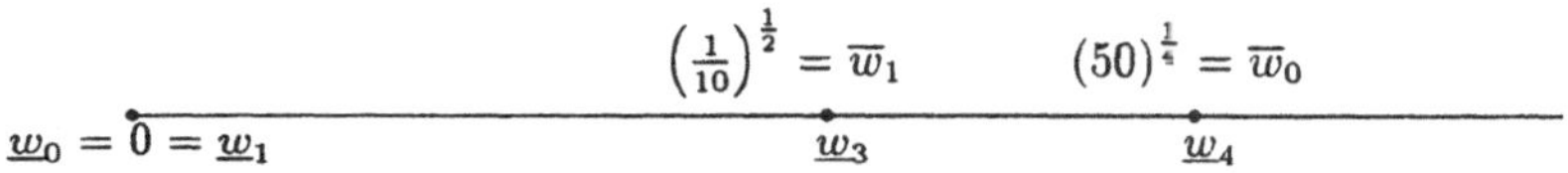

The "interlacing" condition is violated when $w \in \left[\left(\frac{1}{10}\right)^{\frac{1}{2}}, (50)^{\frac{1}{4}}\right]$. Four vertices of the value set $\mathcal{P}_w$ for these w are generated by the polynomials

$$\mathbf{p}_c(s) + 500, \ \mathbf{p}_c(s) - 500, \ \mathbf{p}_c(s) + 500s^3, \ \mathbf{p}_c(s) - 500s^3,$$

where $\mathbf{p}_c(s) = \mathbf{p}(s, \mathbf{0}) = 1000 + 1250s + 1350s^2 + 1000s^3 + 100s^4 + s^5$. The unstable polynomial $\mathbf{p}(s, \mathbf{x}^*) = 0.6\, [\mathbf{p}_c(s) + 500] + 0.4\, [\mathbf{p}_c(s) + 500s^3]$, and the stability is lost exactly when the "interlacing" condition fails.

Chapter 3

Stability Radii and Convex Analysis

This chapter takes care of polynomial families whose coefficients are affine functions of parameters. Specifically, consider polynomials with complex coefficients whose degrees do not exceed n

$$\mathbf{p}(s, \mathbf{d}) = t_0(\mathbf{d}) + t_1(\mathbf{d})s + \ldots + t_n(\mathbf{d})s^n, \tag{3.0.1}$$

where polynomials coefficients are affine functions of the row perturbation vector

$$\mathbf{d} = (a_0 + jb_0, a_1 + jb_1, \ldots, a_{m-1} + jb_{m-1}) \in \mathbf{C}^m,$$

i.e.,

$$(t_0(\mathbf{d}), t_1(\mathbf{d}), \ldots, t_n(\mathbf{d})) = \left(\langle \mathbf{d}, \overline{\mathbf{T}^0} \rangle + t_0^0, \ldots, \langle \mathbf{d}, \overline{\mathbf{T}^n} \rangle + t_n^0\right) = \mathbf{d}T + \mathbf{t}^0. \tag{3.0.2}$$

Here T is an $m \times (n+1)$ matrix with columns $\left[\mathbf{T}^0, \mathbf{T}^1, \ldots, \mathbf{T}^{n-1}, \mathbf{T}^n\right]$, and $\langle \cdot, \cdot \rangle$ stands for the dot product, i.e., for two m dimensional complex vectors $\mathbf{u}$ and $\mathbf{v}$ the dot product is given by $\langle \mathbf{u}, \mathbf{v} \rangle = \sum_{i=1}^{m} u_i \overline{v}_i$. When $\mathbf{d} = \mathbf{0}$ we obtain the nominal polynomial

$$\mathbf{p}(s, \mathbf{0}) = t_0^0 + t_1^0 s + \ldots + t_n^0 s^n$$

of degree n, which is assumed to be Ω stable. Finally, let $f : \mathbf{C}^m \to \mathbf{R}$ be a nonnegative convex function with bounded level sets.

Definition 3.0.2 *The stability radius of the Ω stable nominal polynomial $\mathbf{p}(s, \mathbf{0})$ is defined by*

$$r = \inf\left\{f(\mathbf{d}) \ :\ \mathbf{d} \in \mathbf{C}^m,\ \textit{and}\ \mathbf{p}(s, \mathbf{d})\ \textit{is unstable}\right\}. \tag{3.0.3}$$

The main goal of this chapter is to provide a computationally tractable technique for evaluation of the stability radius r.

The presented framework is general enough to cover many applications where the region Ω can be, for example, the left–half plane, or the open unit disc, and the convex set that contains perturbations $\mathbf{d}$ may have an arbitrary shape, which includes polytopes and ellipsoids as special cases. For further discussion on the framework we refer to [QD]. In order to bridge between different results already available in the literature we concentrate on functions f given by

$$f(\mathbf{d}) = f(a_0 + \boldsymbol{j}b_0, a_1 + \boldsymbol{j}b_1, \ldots, a_{m-1} + \boldsymbol{j}b_{m-1}) = \left[\sum_{k=0}^{m-1} \left|\frac{a_k}{\alpha_k}\right|^p + \sum_{k=0}^{m-1} \left|\frac{b_k}{\beta_k}\right|^p\right]^{1/p}, \quad (3.0.4)$$

where α_k, β_k, $k = 0, 1, \ldots, m-1$ are positive constants, and $p = 1, 2$, or ∞. The function f given by (3.0.4) in general does not satisfy the relation $f((x + \boldsymbol{j}y)\mathbf{d}) = |x + \boldsymbol{j}y| f(\mathbf{d})$. Although we are dealing with families of complex polynomials the stability radius r should not be confused with the complex stability radius introduced by Hinrichsen and Pritchard (see e.g., [HP2]). In order to introduce the complex stability radius r_C one has to consider norms in $\mathbf{C}^m$. For example functions

$$f(\mathbf{d}) = \left[\sum_{k=0}^{m-1} \left|\frac{a_k + \boldsymbol{j}b_k}{\alpha_k}\right|^p\right]^{1/p} = \left[\sum_{k=0}^{m-1} \left|\frac{d_k}{\alpha_k}\right|^p\right]^{1/p} \quad (3.0.5)$$

define "disc" polynomial families, and (3.0.3) with f given by (3.0.5) defines the complex stability radius r_C. If the perturbation $\mathbf{d} \in \mathbf{R}^m$, and the function f is defined in $\mathbf{R}^m$ one can introduce the real stability radius r_R. The corresponding modification of (3.0.4) is given below.

$$f(\mathbf{d}) = f(a_0, a_1, \ldots, a_{m-1}) = \left[\sum_{k=0}^{m-1} \left|\frac{a_k}{\alpha_k}\right|^p\right]^{1/p}. \quad (3.0.6)$$

Our goal is to evaluate the stability radii r, r_C, r_R, and to describe relationships between them.

3.1. Stability Radii and Degree Dropping

A way to determine a stability radius is the following: Denote by $\mathcal{Z}_\gamma$ the set of all roots of the family $\{\mathbf{p}(s, \mathbf{d}),\ f(\mathbf{d}) \le \gamma\}$ (so, for example, $\mathcal{Z}_0$ is the set of all roots of the nominal polynomial $\mathbf{p}(s, \mathbf{0})$). The set $\mathcal{Z}_0$ lies in Ω, when γ is small the "cloud" of roots $\mathcal{Z}_\gamma$ is located in Ω. As γ grows $\mathcal{Z}_\gamma$ grows until it touches $\partial\Omega$ the boundary of the stability region Ω. The minimal γ whose "cloud" of roots $\mathcal{Z}_\gamma$ touches $\partial\Omega$ determines the stability radius. Since the "contact" point $\delta(w) \in \partial\Omega$ is unknown we first evaluate

$$r(w) = \inf_{\mathbf{d}}\{f(\mathbf{d}),\ \mathbf{p}(\delta(w),\mathbf{d}) = 0\} \tag{3.1.1}$$

for each $w \in I_\Omega$. Next a "brute force search" over the boundary of Ω is carried out, i.e., the stability radius is given by $\inf\{r(w),\ w \in I_\Omega\}$.

These arguments can be traced back to the work of Frazer and Duncan [FD], where a family of real polynomials (Im $t_i = 0$, $i = 0,\ldots,n$) with positive leading coefficients (Re $t_n > 0$) is considered. However, as we show by an example, when one deals with polynomials without additional degree assumptions the "degree dropping" may cause the lost of stability.

Example 3.1.1 Let Ω be the left–half plane, $\mathbf{t}^0 = (2,2,1)$, $T = I_{3\times3}$, $\mathbf{d} \in \mathbf{R}^3$, and

$$f(\mathbf{d}) = |\mathbf{d}|_\infty = \max\{|d_1|,\ |d_2|,\ |d_3|\}.$$

The perturbed polynomials are $\mathbf{p}(s,\mathbf{d}) = (2+d_1) + (2+d_2)s + (1+d_3)s^2$, and the nominal polynomial

$$\mathbf{p}(s,\mathbf{0}) = 2 + 2s + s^2 = (s+1+\boldsymbol{j})(s+1-\boldsymbol{j})$$

is stable. Let $w \neq 0$, so if $\mathbf{p}(\boldsymbol{j}w,\mathbf{d}) = 0$ for some $\mathbf{d} \in \mathbf{R}^3$, then

$$\mathbf{p}(s,\mathbf{d}) = \alpha(s-\boldsymbol{j}w)(s+\boldsymbol{j}w) = \alpha[w^2+s^2],\ \text{and}\ f(\mathbf{d}) \geq 2.$$

When $\mathbf{p}(0,\mathbf{d}) = 0$ one has $2 + d_1 = 0$, and $f(\mathbf{d}) \geq 2$. On the other hand it is clear, that the real stability radius can not exceed 1 (and it is easy to show that in fact the real stability radius is exactly 1).

Suppose for a moment that for each unstable polynomial $\mathbf{p}(s,\mathbf{d})$ and each positive ϵ one can find $w \in I_\Omega$ and a perturbation $\mathbf{d}'$ so that

$$\mathbf{p}(\delta(w),\mathbf{d}') = 0,\ \text{and}\ f(\mathbf{d}') \leq f(\mathbf{d}) + \epsilon. \tag{3.1.2}$$

Then

$$\inf_{\mathbf{d}}\{f(\mathbf{d})\ :\ \mathbf{p}(s,\mathbf{d})\ \text{is unstable}\} = \inf_{w}\inf_{\mathbf{d}}\{f(\mathbf{d})\ :\ \mathbf{p}(\delta(w),\mathbf{d}) = 0\}. \tag{3.1.3}$$

We next present conditions under which (3.1.2) holds.

Theorem 3.1.1 Consider the following three conditions:

1. perturbations $\mathbf{d} \in \mathbf{C}^m$,

2. Ω is bounded,

3. Ω^c is bounded. (Here Ω^c stands for the complement of Ω, i.e., $\Omega^c = \mathbf{C} \setminus \Omega^c$.)

Each one of the above conditions implies (3.1.2).

In order to prove the theorem we need auxiliary results given below:

Theorem 3.1.2 Let $\psi_n(s)$ be a sequence of functions, each analytic in a region D bounded by a simple closed contour, and let $\psi_n(s) \to \psi(s)$ uniformly in D. Suppose that $\psi(s)$ is not identically zero. Let s_0 be an interior point of D. Then s_0 is a zero of $\psi(s)$ if, and only if, it is a limit point of the set of zeros of the functions $\psi_n(s)$, points which are zeros for an infinity of values of n being counted as limit-points.

Proof: See [Ti], p. 119, Theorem of Hurwitz.

Lemma 3.1.1 *Consider two polynomials* $\mathbf{p}(s, \mathbf{d}^0)$ *and* $\mathbf{p}(s, \mathbf{d}^1)$. *Suppose that:*

1. $\mathbf{p}(s, \mathbf{d}^0)$ *is stable,*

2. $\mathbf{p}(s, \mathbf{d}^1)$ *is unstable,*

3. deg $\mathbf{p}(s, (1-\lambda)\mathbf{d}^0 + \lambda\mathbf{d}^1) = n$ *for each* $0 \le \lambda \le 1$.

There exists $\lambda_0 \in (0, 1]$, *and* $w_0 \in I_\Omega$ *such that* $\mathbf{p}(\delta(w_0), (1-\lambda_0)\mathbf{d}^0 + \lambda_0\mathbf{d}^1) = 0$.

Proof: The proof follows from the continuous dependence of the roots of $\mathbf{p}(s, (1-\lambda)\mathbf{d}^0 + \lambda\mathbf{d}^1)$ on λ, and the fact that one extreme polynomial is stable, and the other one is unstable.

Lemma 3.1.2 *Let* $\mathbf{p}(s, \mathbf{d})$ *be an unstable polynomial. Then one of the two following conditions holds:*

1. $\exists \lambda_0 \in (0, 1]$ *and* $w_0 \in I_\Omega$ *such that* $\mathbf{p}(\delta(w_0), \lambda_0\mathbf{d}) = 0$.

2. $\exists \lambda_0 \in (0, 1]$ *such that*

(a) deg $\mathbf{p}(s, \lambda_0\mathbf{d}) < n$,

(b) $\forall \lambda \in [0, \lambda_0]$ *the polynomial* $\mathbf{p}(s, \lambda\mathbf{d})$ *is stable.*

(c) $\forall \lambda \in (\lambda_0, 1]$ *the polynomial* $\mathbf{p}(s, \lambda\mathbf{d})$ *is unstable.*

Proof: We assume that Condition 1 fails, and consider the family of polynomials $\{\mathbf{p}(s, \lambda\mathbf{d})\}$, $\lambda \in [0,1]$. The stable polynomial $\mathbf{p}(s, \mathbf{0})$, and the unstable polynomial $\mathbf{p}(s, \mathbf{d})$ are the extreme polynomials of the family. We next show that Conditions 2a, 2b, and 2c hold.

Condition 2a. Suppose the opposite, i.e., deg $\mathbf{p}(s, \lambda\mathbf{d}) = n$ for each $\lambda \in (0,1]$. We apply Lemma 3.1.1, and obtain Condition 1. This contradiction completes the proof of 2a.

Condition 2b. Suppose the opposite, i.e., the polynomial $\mathbf{p}(s, \lambda\mathbf{d})$ is unstable for some $\lambda \in [0, \lambda_0]$. If $\lambda = \lambda_0$, then due to Theorem 3.1.2 there exists λ_1, $0 < \lambda_1 < \lambda_0$ such that $\mathbf{p}(s, \lambda_1\mathbf{d})$ is unstable. So without any loss of generality we can assume that $\lambda < \lambda_0$. An application of Lemma 3.1.1 to polynomials $\mathbf{p}(s, \mathbf{0})$ and $\mathbf{p}(s, \lambda\mathbf{d})$ shows that Condition 1 holds. This contradiction completes the proof of 2b.

Condition 2c. The proof is analogous to that of Condition 2b.

We now proof Theorem 3.1.1. Let $\mathbf{p}(s, \mathbf{d})$ be an unstable polynomial, and $\epsilon > 0$. Consider the polynomial segment $\{\mathbf{p}(s, \lambda\mathbf{d}) \,:\, 0 \leq \lambda \leq 1\}$. If the first condition of Lemma 3.1.2 holds, then $\mathbf{d}' = \lambda_0\mathbf{d}$, and $w = w_0$. We assume that the second condition of the lemma holds and consider three different cases.

Complex perturbations

In this case one has $0 = t_n(\lambda_0\mathbf{d}) = t_n^0 + \langle \lambda_0\mathbf{d}, \overline{\mathbf{T}}^n \rangle = t_n^0 + \lambda_0\langle \mathbf{d}, \overline{\mathbf{T}}^n \rangle$, where $\mathbf{T}^n$ is the last column of the matrix T. This implies that the nonzero two dimensional real vectors $(\text{Re } t_n^0, \text{Im } t_n^0)$ and $(\text{Re }\langle \mathbf{d}, \overline{\mathbf{T}}^n \rangle, \text{Im }\langle \mathbf{d}, \overline{\mathbf{T}}^n \rangle)$ are proportional. Hence for each real $\gamma \neq 0$ the equation

$$t_n^0 + \lambda\langle [1 + \boldsymbol{j}\gamma]\mathbf{d}, \overline{\mathbf{T}}^n \rangle = 0$$

does not have real solutions λ. Due to Theorem 3.1.2 the polynomial $\mathbf{p}(s, \lambda_0[1+\boldsymbol{j}\gamma]\mathbf{d})$ is unstable when γ is small, on the other hand deg $\mathbf{p}(s, \lambda[1 + \boldsymbol{j}\gamma]\mathbf{d}) = n$ for each $\lambda \in [0,1]$. We apply Lemma 3.1.1 to polynomials $\mathbf{p}(s, \mathbf{0})$ and $\mathbf{p}(s, \lambda_0[1 + \boldsymbol{j}\gamma]\mathbf{d})$ and obtain $\mu \in (0, \lambda_0]$, and $w \in I_\Omega$ such that $\mathbf{p}(\delta(w), \mu[1 + \boldsymbol{j}\gamma]\mathbf{d}) = 0$. Hence for small γ one has

$$f(\mu[1 + \boldsymbol{j}\gamma]\mathbf{d}) \leq f([1 + \boldsymbol{j}\gamma]\mathbf{d}) < f(\mathbf{d}) + \epsilon.$$

Real perturbations, bounded Ω

In this case $\forall \mu \in [0, \lambda_0)$, one has $\mathbf{p}(s, \mu\mathbf{d}) = \sigma_\mu(s - s_{\mu 1})\dots(s - s_{\mu n})$, $\lim_{\mu \to \lambda_0} \sigma_\mu = 0$, and $\mathbf{p}(s, \lambda_0\mathbf{d}) = \lim_{\mu \to \lambda_0} \mathbf{p}(s, \mu\mathbf{d})$ for each complex s. Since Ω is bounded, and $\lambda_0 \neq 0$

$$\lambda_0\mathbf{p}(s, \mathbf{d}) + (1 - \lambda_0)\mathbf{p}(s, \mathbf{0}) = \mathbf{p}(s, \lambda_0\mathbf{d}) = \lim_{\mu \to \lambda_0} \mathbf{p}(s, \mu\mathbf{d}) = \lim_{\mu \to \lambda_0} \sigma_\mu \prod_{i=1}^{n} (s - s_{\mu i}) = 0.$$

In other words

$$\mathbf{p}(s,\mathbf{d}) = \frac{\lambda_0 - 1}{\lambda_0}\mathbf{p}(s,\mathbf{0}).$$

The last equation shows that $\mathbf{p}(s,\mathbf{d})$ is stable, contradicts the assumption, and completes the proof.

Real perturbations, bounded Ω^c

Let $s_\mu \in \Omega^c$ be an unstable root of the unstable polynomial $\mathbf{p}(s,\mu\mathbf{d})$, $\lambda_0 < \mu \leq 1$. Let s_0 be a cluster point of the set $\{s_\mu\}$, $\lambda_0 < \mu \leq 1$. Now we abuse the notations and denote by $\{s_\mu\}$ a sequence that converges to s_0. Since Ω^c is a closed set $s_0 \in \Omega^c$. Furthermore,

$$\lim_{\mu\downarrow\lambda_0} |\mathbf{p}(s_0,\mu\mathbf{d})| = \lim_{\mu\downarrow\lambda_0} |\mathbf{p}(s_0,\mu\mathbf{d}) - \mathbf{p}(s_\mu,\mu\mathbf{d})| =$$

$$\lim_{\mu\downarrow\lambda_0} \left|\sum_{k=0}^{n} \langle\mu\mathbf{d},\overline{\mathbf{T}^k}\rangle \left(s_0^k - s_\mu^k\right)\right| \leq \lim_{\mu\downarrow\lambda_0} \sum_{k=0}^{n} \left|\langle\mathbf{d},\overline{\mathbf{T}^k}\rangle\right| \cdot \left|s_0^k - s_\mu^k\right| = 0,$$

and

$$0 = \lim_{\mu\downarrow\lambda_0} \mathbf{p}(s_0,\mu\mathbf{d}) = \mathbf{p}(s_0,\lambda_0\mathbf{d}) \neq 0.$$

This contradiction completes the proof.

The next theorem summarizes results presented so far.

Theorem 3.1.3 If at least one of the conditions listed in Theorem 3.1.1 is satisfied, then

$$\inf_{\mathbf{d}} \{f(\mathbf{d}) \ : \ \mathbf{p}(s,\mathbf{d}) \text{ is unstable}\} = \inf_{w}\inf_{\mathbf{d}} \{f(\mathbf{d}) \ : \ \mathbf{p}(\delta(w),\mathbf{d}) = 0\}.$$

Finally we address the case motivated by Example 3.1.1: real perturbations, and unbounded regions Ω and Ω^c .

Theorem 3.1.4 Suppose that perturbations $\mathbf{d} \in \mathbf{R}^m$, T is a matrix with real entries, and both regions Ω, and Ω^c are unbounded. If Ω^c contains real numbers of an arbitrary large magnitude, then

$$r_R = \inf \{f(\mathbf{d}) \ : \ \mathbf{p}(s,\mathbf{d}) \text{ is unstable}\} = \min \{r_1, \ r_\infty\} \qquad (3.1.4)$$

where

$$r_1 = \inf_{w}\inf_{\mathbf{d}} \{f(\mathbf{d}) \ : \ \mathbf{p}(\delta(w),\mathbf{d}) = 0\}, \text{ and } r_\infty = \inf_{\mathbf{d}} \{f(\mathbf{d}) \ : \ t_n(\mathbf{d}) = 0\}.$$

The proof of the theorem follows from the lemma given next.

Lemma 3.1.3 *Suppose that for some* $\mathbf{d} \in \mathbf{R}^m$ *the degree of the corresponding polynomial* $\mathbf{p}(s, \mathbf{d})$ *does not exceed* $n-1$, *i.e.,* $\mathbf{p}(s, \mathbf{d}) = t_0(\mathbf{d}) + t_1(\mathbf{d})s + \ldots + t_{n-1}(\mathbf{d})s^{n-1}$. *Then for each positive integer* l *there exist* $s_l \in \Omega^c$, *and* $\mathbf{d}^l \in \mathbf{R}^m$ *such that*

1. $f(\mathbf{d}^l) < \frac{1}{l}$.
2. $\mathbf{p}(s_l, \mathbf{d} + \mathbf{d}^l) = 0$.

Proof:

The conditions $t_n(\mathbf{d}) = t_n^0 + \langle \mathbf{d}, \mathbf{T}^n \rangle = 0$, and $t_n^0 \neq 0$ imply $\langle \mathbf{T}^n, \mathbf{T}^n \rangle \neq 0$. Let

$$\mu_i = \frac{\langle \mathbf{T}^i, \mathbf{T}^n \rangle}{\langle \mathbf{T}^n, \mathbf{T}^n \rangle}, \qquad i = 0, \ldots, n-1.$$

In the expression

$$-\frac{\mathbf{p}(s, \mathbf{d})}{\langle \mathbf{T}^n, \mathbf{T}^n \rangle (\mu_0 + \mu_1 s + \ldots + \mu_{n-1} s^{n-1} + s^n)}$$

the degree of the numerator is less than the degree of the denominator. One can always pick a real $s_l \in \Omega^c$ such that

$$f\left(-\frac{\mathbf{p}(s_l, \mathbf{d})}{\langle \mathbf{T}^n, \mathbf{T}^n \rangle (\mu_0 + \mu_1 s_l + \ldots + \mu_{n-1} s_l^{n-1} + s_l^n)} \mathbf{T}^n\right) < \frac{1}{l},$$

and

$$-\frac{\mathbf{p}(s_l, \mathbf{d})}{\langle \mathbf{T}^n, \mathbf{T}^n \rangle (\mu_0 + \mu_1 s_l + \ldots + \mu_{n-1} s_l^{n-1} + s_l^n)} \neq 0.$$

Let

$$k_l = -\frac{\mathbf{p}(s_l, \mathbf{d})}{\langle \mathbf{T}^n, \mathbf{T}^n \rangle (\mu_0 + \mu_1 s_l + \ldots + \mu_{n-1} s_l^{n-1} + s_l^n)}, \qquad \text{and} \qquad \mathbf{d}^l = k_l \mathbf{T}^n.$$

Then

$$\mathbf{p}(s_l, \mathbf{d} + \mathbf{d}^l) = \mathbf{p}(s_l, \mathbf{d} + k_l \mathbf{T}^n) = \mathbf{p}(s_l, \mathbf{d}) + \sum_{i=0}^{n-1} \langle \mathbf{T}^i, k_l \mathbf{T}^n \rangle s_l^i + \langle \mathbf{T}^n, k_l \mathbf{T}^n \rangle s_l^n =$$

$$\mathbf{p}(s_l, \mathbf{d}) + k_l \langle \mathbf{T}^n, \mathbf{T}^n \rangle \left[\sum_{i=0}^{n-1} \mu_i s_l^i + s_l^n \right] = 0.$$

On the other hand,

$$f(\mathbf{d}^l) = f(k_l \mathbf{T}^n) = f\left(-\frac{\mathbf{p}(s_l, \mathbf{d})}{\langle \mathbf{T}^n, \mathbf{T}^n \rangle (\mu_0 + \mu_1 s_l + \ldots + \mu_{n-1} s_l^{n-1} + s_l^n)} \mathbf{T}^n\right) < \frac{1}{l}.$$

The lemma shows that the real stability radius cannot exceed r_∞. If $f(\mathbf{d}) < r_\infty$ and $\mathbf{p}(s, \mathbf{d})$ is unstable, then $\mathbf{p}(\delta(w), \lambda \mathbf{d}) = 0$ for some $\lambda \in (0, 1]$, and $w \in I_\Omega$. This implies $r_1 \leq f(\mathbf{d})$, and completes the proof of the theorem.

3.2. Constrained Optimization Problem

Evaluation of $r(w) = \inf_{\mathbf{d}} \{f(\mathbf{d}) \ : \ \mathbf{p}(\delta(w), \mathbf{d}) = 0\}$ is the main computational burden associated with robust stability problems. In this section we follow the development of Qiu and Davison [QD], and reduce the computation of $r(w)$ to a solution of a constrained optimization problem. The latter will be solved in the next two sections.

The constraints of the problem $\mathbf{p}(\delta(w), \mathbf{d}) = 0$ can be written as

$$\mathbf{p}(\delta(w), \mathbf{d}) - \mathbf{p}(\delta(w), \mathbf{0}) = -\mathbf{p}(\delta(w), \mathbf{0}). \tag{3.2.1}$$

Let $\boldsymbol{\delta}(w) = (1, \delta(w), \ldots, \delta^n(w))^t$. Equation (3.2.1) can be written now as

$$\langle \mathbf{d}T, \overline{\boldsymbol{\delta}(w)} \rangle = -\mathbf{p}(\delta(w), \mathbf{0}). \tag{3.2.2}$$

Since the nominal polynomial $\mathbf{p}(s, \mathbf{0})$ has no roots on the boundary of Ω one has $\mathbf{p}(\delta(w), \mathbf{0}) \neq 0$, and equation (3.2.2) becomes

$$\langle \mathbf{d}, \mathbf{s}(w) \rangle = 1, \text{ where } \mathbf{s}(w) = -T \overline{\left[\frac{\boldsymbol{\delta}(w)}{\mathbf{p}(\delta(w), \mathbf{0})} \right]}. \tag{3.2.3}$$

Thus $r(w)$ is given by

$$r(w) = \inf \{ f(\mathbf{d}) : \mathbf{d} \in \mathbf{C}^m, \ \langle \mathbf{d}, \mathbf{s}(w) \rangle = 1 \}. \tag{3.2.4}$$

By definition $r(w) = \infty$ if and only if there exists no $\mathbf{d} \in \mathbf{C}^m$ such that $\langle \mathbf{d}, \mathbf{s}(w) \rangle = 1$ (i.e., $\mathbf{s}(w) = \mathbf{0}$). In the next section the constrained optimization problem is solved for the special case when the function f is a norm in $\mathbf{C}^m$.

3.3. Complex Stability Radius

In this section we consider polynomial families (3.0.1) with complex perturbations. In addition we assume that the function f is a norm which we denote by $||\cdot||$. According to Theorem 3.1.3

$$r_C = \inf \{ r(w) \ : \ w \in I_\Omega \}, \text{ where } r(w) = \inf_{\mathbf{d}} \{ ||\mathbf{d}||, \ \mathbf{d} \in \mathbf{C}^m, \ \langle \mathbf{d}, \mathbf{s}(w) \rangle = 1 \}. \tag{3.3.1}$$

The mapping $\Lambda \ : \ \mathbf{C}^m \to \mathbf{C}$ defined by $\Lambda(\mathbf{d}) = \langle \mathbf{d}, \mathbf{s}(w) \rangle = \sum_{i=0}^{m-1} d_i \overline{s_i(w)}$ is a continuous linear functional. If $1 = \langle \mathbf{d}, \mathbf{s}(w) \rangle$, then

$$1 = |\langle \mathbf{d}, \mathbf{s}(w) \rangle| = \left| \sum_{i=0}^{m-1} d_i \overline{s_i(w)} \right| \leq ||\mathbf{d}|| \cdot ||\Lambda||.$$

where $||\cdot||_*$ is the dual norm, i.e., $||\Lambda||_* = \sup_{||\mathbf{d}||\neq 0} \frac{|\Lambda(\mathbf{d})|}{||\mathbf{d}||}$. This shows that $\frac{1}{||\Lambda||_*} \leq ||\mathbf{d}||$. On the other hand, if $||\overline{\mathbf{s}(w)}||_* \neq 0$, according to the Hahn–Banach Theorem there exists $\mathbf{D} \in \mathbf{C}^m$ such that

$$1 = \langle \mathbf{D}, \mathbf{s}(w) \rangle = ||\mathbf{D}|| \cdot ||\Lambda||_*, \text{ and } ||\mathbf{D}|| = \frac{1}{||\Lambda||_*}.$$

Keeping in mind that $||\Lambda||_* = ||\overline{\mathbf{s}(w)}||_*$ we obtain the following result.

Theorem 3.3.1 The complex stability radius r_C is given by

$$r_C = \inf \{ r(w) \ : \ w \in I_\Omega \},$$

where

$$r(w) = \begin{cases} \infty & \text{if } \left\| T \left[\frac{\delta(w)}{\mathbf{p}(\delta(w),\mathbf{0})} \right] \right\|_* = 0 \\ \left[\left\| T \left[\frac{\delta(w)}{\mathbf{p}(\delta(w),\mathbf{0})} \right] \right\|_* \right]^{-1} & \text{otherwise.} \end{cases}$$

In particular, when f is the l_p norm given by (3.0.5), due to Hölder's inequality the dual norm $||\cdot||_*$ is the l_q weighted norm given by

$$||\mathbf{D}||_* = \left[\sum_{i=0}^{m-1} |\alpha_i D_i|^q \right]^{1/q}, \quad p + q = pq. \tag{3.3.2}$$

We next consider examples of Hurwitz and Schur stability for polynomial families centered at a stable polynomial

$$\mathbf{p}(s, \mathbf{0}) = (a_0^0 + jb_0^0) + (a_1^0 + jb_1^0)s + \ldots + (a_n^0 + jb_n^0)s^n.$$

We assume that $m = n+1$, the matrix T is the identity matrix $I_{(n+1)\times(n+1)}$, and f is given by (3.0.5). Let

$$\mathbf{p}(\delta(w), \mathbf{0}) = U(w) + jV(w). \tag{3.3.3}$$

Example 3.3.1 Hurwitz robust l_p stability. In this case

$$U(w) = a_0^0 - b_1^0 w - a_2^0 w^2 + b_3^0 w^3 + \ldots, \text{ and } V(w) = b_0^0 + a_1^0 w - b_2^0 w^2 - a_3^0 w^3 + \ldots.$$

$$r(w) = \frac{|U(w) + jV(w)|}{[\sum_{k=0}^n |\alpha_k w^k|^q]^{\frac{1}{q}}}, \text{ and } r_C = \inf \{ r(w) \ : \ -\infty < w < \infty \}. \tag{3.3.4}$$

Hurwitz l_p stability of monic polynomials with complex coefficients has been investigated by Chapellat, Bhattacharyya and Dahleh [CBD], Desages, Castro and Cendra [DDC], Polyak and Tsypkin [PT2], and Li, Nagpal, and Lee [LNL].

Example 3.3.2 Schur robust l_p stability.
In this case

$$U(w) = \sum_{k=0}^{n} \left[a_k^0 \cos kw - b_k^0 \sin kw\right], \text{ and } V(w) = \sum_{k=0}^{n} \left[a_k^0 \sin kw + b_k^0 \cos kw\right].$$

$$r(w) = \frac{|U(w) + jV(w)|}{\left[\sum_{k=0}^{n} |\alpha_k|^q\right]^{\frac{1}{q}}}, \text{ and } r_C = \inf\{r(w) \ : \ 0 \le w < 2\pi\}. \tag{3.3.5}$$

Investigation of robust stability for discrete-time systems has been conducted by Polyak and Tsypkin [PT2].

3.4. Stability Radius ,

In this section we evaluate the stability radius r defined by (3.0.3). Our analysis is based on the results of Teboulle and Kogan [TK]. In order to compute $r(w)$ one has to solve the optimization problem (3.2.4)

$$r(w) = \inf\{f(\mathbf{d}) : \mathbf{d} \in \mathbf{C}^m, \ \langle \mathbf{d}, \mathbf{s}(w)\rangle = 1\}.$$

To avoid trivial situations we assume that $\mathbf{s}(w) \neq 0$.

Let $\mathbf{s}(w) = (\phi_0 + j\psi_0, \ldots, \phi_{m-1} + j\psi_{m-1})^t$. We identify the complex m dimensional space $\mathbf{C}^m$ with the real space $\mathbf{R}^{2m}$, slightly abuse notations, and solve the convex minimization problem with two linear constraints:

$$r(w) = \inf_{\mathbf{a},\,\mathbf{b}} \left\{ f(\mathbf{a} + j\mathbf{b}) \ : \ \sum_{k=0}^{m-1} a_k\phi_k + b_k\psi_k = 1, \ \sum_{k=0}^{m-1} b_k\phi_k - a_k\psi_k = 0 \right\}. \tag{3.4.1}$$

Auxiliary convex duality results that reduce the constrained minimization problem (3.4.1) to an unconstrained optimization problem in $\mathbf{R}^2$ are introduced next.

3.4.1. Convex Duality Results

Let $h \ : \ \mathbf{R}^k \to \mathbf{R}$ be a convex function. Consider a constrained minimization problem in $\mathbf{R}^k$

$$V = \inf\left\{h(\mathbf{x}) \ : \ \mathbf{x} \in \mathbf{R}^k \text{ subject to } \langle \mathbf{c}^i, \mathbf{x}\rangle = \nu_i, \ i = 1, \ldots, l\right\}. \tag{3.4.2}$$

If the problem (3.4.2) is feasible, i.e., there exists $\mathbf{x}$ so that $\langle \mathbf{c}^i, \mathbf{x}\rangle = \nu_i$, $i = 1, \ldots, l$, then

$$\begin{aligned} V &= \sup_{\boldsymbol{\lambda}} \inf_{\mathbf{x}} \left\{ h(\mathbf{x}) + \sum_{i=1}^{l} \lambda_i \left(\nu_i - \langle \mathbf{c}^i, \mathbf{x}\rangle\right)\right\} \\ &= \sup_{\boldsymbol{\lambda}} \left\{ \sum_{i=1}^{l} \lambda_i\nu_i - \sup_{\mathbf{x}} \left(\langle \sum_{i=1}^{l} \lambda_i \mathbf{c}^i, \mathbf{x}\rangle - h(\mathbf{x})\right)\right\} \end{aligned} \tag{3.4.3}$$

where $\boldsymbol{\lambda}$ s a vector of Lagrange multipliers (see e.g., [Ro]).

Definition 3.4.1 *For a given function* $h : \mathbf{R}^k \to \mathbf{R}$ *its conjugate* $h^* : \mathbf{R}^k \to \mathbf{R}$ *is defined by*

$$h^*(\mathbf{y}) = \sup_{\mathbf{x}} \{\langle \mathbf{y}, \mathbf{x} \rangle - h(\mathbf{x})\}.$$

The relation (3.4.3) can now be written as

$$V = \sup_{\boldsymbol{\lambda}} \left\{ \sum_{i=1}^{l} \lambda_i \nu_i - h^* \left(\sum_{i=1}^{l} \lambda_i \mathbf{c}^i \right) \right\}. \tag{3.4.4}$$

Motivated by (3.4.1) we are primarily interested in a special case of the minimization problem with two constraints, where $\nu_1 = 1$, and $\nu_2 = 0$. We denote the two Lagrange multipliers by λ, and μ and obtain

$$V = \sup_{\lambda, \mu} \lambda - h^*(\lambda \mathbf{c}^1 + \mu \mathbf{c}^2). \tag{3.4.5}$$

If, in addition, $h(\mathbf{x})$ is positively homogeneous and satisfies $h(\mathbf{0}) = 0$, i.e., $h(a\mathbf{x}) = ah(\mathbf{x})$ for $a \geq 0$, then the expression (3.4.5) can be further simplified once the concept of polar function is introduced.

Definition 3.4.2 *The polar* h^0 *of the function* h *is defined by*

$$h^0(\mathbf{y}) = \sup \{\langle \mathbf{x}, \mathbf{y} \rangle \ : \ h(\mathbf{x}) \leq 1\}.$$

Lemma 3.4.1 *If, in addition,* $h(a\mathbf{x}) = ah(\mathbf{x})$ *for* $a \geq 0$, *then* $V = \sup_{\gamma \in \mathbf{R}} \left\{ \frac{1}{h^0(\mathbf{c}^1 + \gamma \mathbf{c}^2)} \right\}$.

Proof: The definitions of conjugate and polar functions imply that

$$h^*(\mathbf{y}) = \begin{cases} 0 & \text{if} \quad h^0(\mathbf{y}) \leq 1 \\ +\infty & \text{otherwise.} \end{cases}$$

Hence

$$\begin{aligned} V &= \sup_{\lambda, \mu} \lambda - h^*(\lambda \mathbf{c}^1 + \mu \mathbf{c}^2) \\ &= \sup \left\{ \lambda \ : \ h^0(\lambda \mathbf{c}^1 + \mu \mathbf{c}^2) \leq 1, \ \mu \in \mathbf{R}, \ \lambda \in \mathbf{R} \right\} \\ &= \sup \left\{ \lambda \ : \ h^0(\lambda \mathbf{c}^1 + \mu \mathbf{c}^2) \leq 1, \ \mu \in \mathbf{R}, \ \lambda > 0 \right\} \\ &= \sup \left\{ \lambda \ : \ \lambda h^0(\mathbf{c}^1 + \frac{\mu}{\lambda} \mathbf{c}^2) \leq 1, \ \mu \in \mathbf{R}, \ \lambda > 0 \right\} \\ &= \frac{1}{\inf \{h^0(\mathbf{c}^1 + \gamma \mathbf{c}^2) \ : \ \gamma \in \mathbf{R}\}} \\ &= \sup_{\gamma \in \mathbf{R}} \left\{ \frac{1}{h^0(\mathbf{c}^1 + \gamma \mathbf{c}^2)} \right\}. \end{aligned}$$

Conjugate and polar functions can be explicitly evaluated for many functions h useful for applications. Below we provide formulae for two special important classes of functions h. The results stated below follow from Hölder's inequality with $p + q = pq$.

Example 3.4.1 *Real weighted l_p norms.*

$$h(\mathbf{x}) = \left[\sum_{i=1}^{k}\left|\frac{x_i}{\chi_i}\right|^p\right]^{\frac{1}{p}}, \ 1 \le p \le \infty, \ 0 < \chi_i, \ i = 1, \ldots, k.$$

$$h^0(\mathbf{y}) = \left[\sum_{i=1}^{k}|\chi_i y_i|^q\right]^{\frac{1}{q}}.$$

$$h^*(\mathbf{y}) = \begin{cases} 0 & \textit{if} \ \left[\sum_{i=1}^{k}|\chi_i y_i|^q\right]^{\frac{1}{q}} \le 1 \\ +\infty & \textit{if} \ \left[\sum_{i=1}^{k}|\chi_i y_i|^q\right]^{\frac{1}{q}} > 1. \end{cases}$$

Finally we note that when $1 < p < \infty$, and x, y are scalars one has

$$\sup_x \left\{ xy - \frac{1}{p}\left|\frac{x}{\chi}\right|^p \right\} = \frac{1}{q}|\chi y|^q. \tag{3.4.6}$$

3.4.2. Dual Optimization Problem

The minimization problem (3.4.1) can now be written as

$$r(w) = \sup_{\lambda,\mu} \{\lambda - f^*(\lambda \mathbf{u} + \mu \mathbf{v})\}, \tag{3.4.7}$$

where

$$\begin{aligned} \mathbf{u} &= (\ \phi_0, \ \ldots, \ \phi_{m-1}, \ \psi_0, \ldots, \psi_{m-1}) \\ \mathbf{v} &= (-\psi_0, \ \ldots, \ -\psi_{m-1}, \ \phi_0, \ldots, \phi_{m-1}). \end{aligned}$$

Thus the original optimization problem (3.4.1) has been reduced to a two dimensional unconstrained optimization problem (3.4.7). Efficient numerical algorithms can be applied to evaluate $r(w)$, if the problem (3.4.7) does not admit analytic solutions. We next turn to the special cases of l_p norms, and provide formulae for $r(w)$.

3.4.3. Special Cases: l_p Norms

When the function f is a weighted l_p norm due to Lemma 3.4.1 and (3.4.7)

$$r(w) = \sup_{\lambda,\mu} \{\lambda - f^*(\lambda \mathbf{u} + \mu \mathbf{v})\} = \sup_{\gamma \in \mathbf{R}} \left\{ \frac{1}{f^0(\mathbf{u} + \gamma \mathbf{v})} \right\} = \left[\inf_{\gamma \in \mathbf{R}} f^0(\mathbf{u} + \gamma \mathbf{v}) \right]^{-1}.$$

The analytic expressions for f^0 are given in Example (3.4.1), and

$$r(w) = \left[\inf_{\gamma} \left[\sum_{i=0}^{m-1} |\alpha_i(u_i + \gamma v_i)|^q + \sum_{i=0}^{m-1} |\beta_i(u_{i+m} + \gamma v_{i+m})|^q \right]^{\frac{1}{q}} \right]^{-1}.$$

The analytic expressions for $r(w)$ are given, for example, in [QD] for the three special cases $p=\infty$, $p=2$, $p=1$

$$r(w)=\begin{cases} \infty & \text{if } \mathbf{v}=\mathbf{u}=\mathbf{0}, \\ [f^0(\mathbf{u})]^{-1} & \text{if } \mathbf{v}=\mathbf{0},\ \mathbf{u}\neq\mathbf{0}, \\ \max\limits_{q,v_q\neq 0}\left[\sum\limits_{i=0}^{m-1}\chi_i\left|\frac{u_iv_q-u_qv_i}{v_q}\right|\right]^{-1} & \text{if } \mathbf{v}\neq\mathbf{0} \text{ and } p=\infty. \\ \left[\dfrac{\sum\limits_{i=0}^{m-1}\chi_i^2v_i^2}{\sum\limits_{i=0}^{m-1}\chi_i^2u_i^2\sum\limits_{i=0}^{m-1}\chi_i^2v_i^2-\left(\sum\limits_{i=0}^{m-1}\chi_i^2u_iv_i\right)^2}\right]^{\frac{1}{2}} & \text{if } \mathbf{v}\neq\mathbf{0}, \text{ and } p=2. \\ \left[\begin{matrix}\min\left|\dfrac{\chi_i\chi_q(u_iv_q-u_qv_i)}{\chi_qv_q+(-1)^l\chi_iv_i}\right| \\ \\ \chi_qv_q+(-1)^l\chi_iv_i\neq 0, \\ l=0,1\end{matrix}\right]^{-1} & \text{if } \mathbf{v}\neq\mathbf{0}, \text{ and } p=1.\end{cases} \tag{3.4.8}$$

Here $\chi_i=\alpha_i$, and $\chi_{m+i}=\beta_i$, $i=0,\ldots,m-1$.

3.5. Examples: Stability Radii for Schur and Hurwitz Polynomials

Throughout the section we assume that $m=n+1$, T is the identity matrix, the functions f are l_p norms given by (3.0.4), and the polynomial families are centered at the stable nominal polynomial

$$\mathbf{p}(s,\mathbf{0})=(a_0^0+\mathbf{j}b_0^0)+(a_1^0+\mathbf{j}b_1^0)s+\ldots+(a_n^0+\mathbf{j}b_n^0)s^n.$$

The values of the nominal polynomial on the boundary of the stability region $\mathbf{p}(\delta(w),\mathbf{0})$ are denoted by $U(w)+\mathbf{j}V(w)$. First we consider the Schur stability case.

Example 3.5.1 l_∞ stability radius of complex Schur polynomials .
The stability region Ω is the open unit disc, and the parameterization of the boundary is given by $\delta(w)=e^{\mathbf{j}w}$, $0\le w<2\pi$. The measure of the perturbation is the weighted l_∞ norm (3.0.4). The vector $\mathbf{s}(w)$ (see 3.2.3) is given by the formula

$$\mathbf{s}=-\overline{\left[\frac{(1,e^{\mathbf{j}w},e^{\mathbf{j}2w},\ldots,e^{\mathbf{j}nw})^t}{U(w)+\mathbf{j}V(w)}\right]}=-\frac{(1,e^{-\mathbf{j}w},e^{-\mathbf{j}2w},\ldots,e^{-\mathbf{j}nw})^t(U(w)+\mathbf{j}V(w))}{U^2(w)+V^2(w)}.$$

So the coordinates $\phi_i + j\psi_i$, $i = 0, \dots n$ of $\mathbf{s}$ are

$$\phi_i = -\frac{U(w)\cos iw + V(w)\sin iw}{U^2(w) + V^2(w)}, \text{ and } \psi_i = -\frac{V(w)\cos iw - U(w)\sin iw}{U^2(w) + V^2(w)}.$$

The vector $[U^2(w) + V^2(w)]\mathbf{u}$ is

$$(-\alpha_0 U(w),\ -\alpha_1[U(w)\cos w + V(w)\sin w],\ \dots,\ -\alpha_n[U(w)\cos nw + V(w)\sin nw],$$
$$-\beta_0 V(w),\ -\beta_1[V(w)\cos w - U(w)\sin w],\ \dots,\ -\beta_n[V(w)\cos nw - U(w)\sin nw]),$$

and the vector $[U^2(w) + V^2(w)]\mathbf{v}$ is

$$(\alpha_0 V(w),\ \alpha_1[V(w)\cos w - U(w)\sin w],\ \dots,\ \alpha_n[V(w)\cos nw - U(w)\sin nw],$$
$$-\beta_0 U(w),\ -\beta_1[U(w)\cos w + V(w)\sin w],\ \dots,\ -\beta_n[U(w)\cos nw + V(w)\sin nw]).$$

The functions $U(w)$ and $V(w)$ can not vanish simultaneously, hence the vector $\mathbf{v}$ does not vanish. If $v_k \neq 0$ for some k, $0 \leq k \leq n$, then

$$\sum_{i=0}^{2n+1} \left| \frac{u_i v_k - u_k v_i}{v_k} \right| = \frac{\sum_{i=0}^{n} \alpha_i |\sin(k-i)w| + \beta_i |\cos(k-i)w|}{|V(w)\cos kw - U(w)\sin kw|}.$$

If $v_k \neq 0$ for some k, $n+1 \leq k \leq 2n$, then

$$\sum_{i=0}^{2n+1} \left| \frac{u_i v_k - u_k v_i}{v_k} \right| = \frac{\sum_{i=0}^{n} \alpha_i |\cos(k-i)w| + \beta_i |\sin(k-i)w|}{|U(w)\cos kw + V(w)\sin kw|}.$$

Let

$$\begin{aligned} x_\infty(w) &= \max_{0 \leq k \leq n} \frac{|V(w)\cos kw - U(w)\sin kw|}{\sum_{i=0}^{n} \alpha_i |\sin(k-i)w| + \beta_i |\cos(k-i)w|} \\ y_\infty(w) &= \max_{n+1 \leq k \leq 2n+1} \frac{|U(w)\cos kw + V(w)\sin kw|}{\sum_{i=0}^{n} \alpha_i |\cos(k-i)w| + \beta_i |\sin(k-i)w|}, \end{aligned}$$

then

$$r(w) = \max\{x_\infty(w),\ y_\infty(w)\}. \tag{3.5.1}$$

Example 3.5.2 l_p stability radius of complex Hurwitz polynomials .
When $p = 1, 2$ or ∞ the expression for $r(w)$ is provided by (3.4.8). Below we evaluate $r(w)$ for each $1 < p < \infty$. Let

$U(w; \mathbf{a}, \mathbf{b}) = a_0 - b_1 w - a_2 w^2 + b_3 w^3 + \dots$, and $V(w; \mathbf{a}, \mathbf{b}) = b_0 + a_1 w - b_2 w^2 - a_3 w^3 + \dots$,

then $r^p(w)$ is the value of the constrained minimization problem

$$\begin{cases} \inf \left[\sum_{i=0}^{n} \left| \frac{a_i - a_i^0}{\alpha_i} \right|^p + \sum_{i=0}^{n} \left| \frac{b_i - b_i^0}{\beta_i} \right|^p \right] \\ \text{subject to} \\ U(w; \mathbf{a}, \mathbf{b}) - U(w) = -U(w),\ V(w; \mathbf{a}, \mathbf{b}) - V(w) = -V(w). \end{cases}$$

Due to the separable structure the problem breaks down into the two problems

$$\begin{cases} \inf\left[\left|\frac{a_0-a_0^0}{\alpha_0}\right|^p+\left|\frac{b_1-b_1^0}{\beta_1}\right|^p+\left|\frac{a_2-a_2^0}{\alpha_2}\right|^p+\left|\frac{b_3-b_3^0}{\beta_3}\right|^p+\dots\right] \\ \text{subject to} \\ (a_0-a_0^0)-(b_1-b_1^0)w-(a_2-a_2^0)w^2+(b_3-b_3^0)w^3+\dots=-U(w), \end{cases}$$

and

$$\begin{cases} \inf\left[\left|\frac{b_0-b_0^0}{\beta_0}\right|^p+\left|\frac{a_1-a_1^0}{\alpha_1}\right|^p+\left|\frac{b_2-b_2^0}{\beta_2}\right|^p+\left|\frac{a_3-a_3^0}{\alpha_3}\right|^p+\dots\right] \\ \text{subject to} \\ (b_0-b_0^0)+(a_1-a_1^0)w-(b_2-b_2^0)w^2-(a_3-a_3^0)w^3+\dots=-V(w). \end{cases}$$

Let V_R be the value of the first problem, and V_I be the value of the second one, then

$$r(w)=[V_R+V_I]^{\frac{1}{p}}.$$

We next evaluate V_R, evaluation of V_I is analogous. Note that $V_R^{\frac{1}{p}}$ is the value of the optimization problem

$$\begin{cases} \inf\left[\left|\frac{a_0-a_0^0}{\alpha_0}\right|^p+\left|\frac{b_1-b_1^0}{\beta_1}\right|^p+\left|\frac{a_2-a_2^0}{\alpha_2}\right|^p+\left|\frac{b_3-b_3^0}{\beta_3}\right|^p+\dots\right]^{\frac{1}{p}} \\ \text{subject to} \\ (a_0-a_0^0)-(b_1-b_1^0)w-(a_2-a_2^0)w^2+(b_3-b_3^0)w^3+\dots=-U(w), \end{cases}$$

Let

$$h(x_0,\dots,x_n)=\frac{1}{p}\left[\left|\frac{x_0}{\alpha_0}\right|^p+\left|\frac{x_1}{\beta_1}\right|^p+\left|\frac{x_2}{\alpha_2}\right|^p+\left|\frac{x_3}{\beta_3}\right|^p+\dots\right]^{\frac{1}{p}},$$

$$\mathbf{v}=-\left(a_0-a_0^0,-(b_1-b_1^0),-(a_2-a_2^0),b_3-b_3^0,\dots\right)^t,\ \text{and}\ \mathbf{w}=\left(1,w,w^2,\dots,w^n\right)^t.$$

Then

$$V_R^{\frac{1}{p}}=\inf_{\mathbf{v}}\{h(\mathbf{v})\ \text{subject to}\ \langle\mathbf{v},\mathbf{w}\rangle=U(w)\}. \tag{3.5.2}$$

This constrained minimization problem is analogous to the problem (3.3.1), and

$$V_R^{\frac{1}{p}}=\begin{cases} 0 & \text{if}\quad U(w)=0 \\ \left\|\frac{\mathbf{w}}{U(w)}\right\|_*^{-1} & \text{otherwise.} \end{cases}$$

According to Hölder's inequality

$$\left\|\frac{\mathbf{w}}{U(w)}\right\|_*=\frac{[|\alpha_0|^q+|\beta_1w|^q+|\alpha_2w^2|^q+|\beta_3w^3|^q+\dots]^{\frac{1}{q}}}{|U(w)|}.$$

Let $S_p(w) = \left[|\alpha_0|^q + |\beta_1 w|^q + |\alpha_2 w^2|^q + |\beta_3 w^3|^q + \ldots\right]^{\frac{1}{q}}$. Then $V_R^{\frac{1}{p}} = \frac{|U(w)|}{S_p(w)}$, and $V_R = \left|\frac{U(w)}{S_p(w)}\right|^p$. Similar calculations show that the value of the second problem V_I is given by $\left|\frac{V(w)}{T_p(w)}\right|^p$, where

$$T_p(w) = \left[|\beta_0|^q + |\alpha_1 w|^q + |\beta_2 w^2|^q + \ldots\right]^{\frac{1}{q}}.$$

Finally let

$$x_p(w) = \frac{U(w)}{S_p(w)}, \text{ and } y_p(w) = \frac{V(w)}{T_p(w)}. \tag{3.5.3}$$

Then

$$r(w) = [V_R + V_I]^{\frac{1}{p}} = [|x_p(w)|^p + |y_p(w)|^p]^{\frac{1}{p}}. \tag{3.5.4}$$

This completes the proof of (3.5.4) for $1 < p < \infty$. Furthermore (3.4.8) implies (3.5.4) for the extreme cases $p = 1, \infty$. The stability radius r is given by

$$r = \inf_{w \in \mathbf{R}} [|x_p(w)|^p + |y_p(w)|^p]^{\frac{1}{p}}.$$

The stability radius for the special case $p = \infty$ has been evaluated by Katbab and Jury [KJ2].

3.6. Real Stability Radius: Examples

To evaluate the real stability radius r_R one has to find $r_1 = \inf r(w)$, and r_∞ (if both the stability domain Ω, and its complement Ω^c are unbounded). The formulae for $r(w)$ given in the previous section remain valid in the real case under the formal assumption $\beta_0 = \beta_1 = \ldots = \beta_{m-1} = 0$. In the examples to follow we consider a stable nominal polynomial

$$\mathbf{p}(s, \mathbf{0}) = a_0^0 + a_1^0 s + \ldots + a_n^0 s^n,$$

whose values on the boundary of the stability domain are denoted by $U(w) + \boldsymbol{j} V(w)$, i.e.,

$$\mathbf{p}(\delta(w), \mathbf{0}) = U(w) + \boldsymbol{j} V(w).$$

We assume that $m = n + 1$, T is the identity matrix, and evaluate the l_p real stability radius for families of Schur and Hurwitz polynomials centered at $\mathbf{p}(s, \mathbf{0})$.

Example 3.6.1 l_∞ stability radius of Schur polynomials .

Let

$$x_\infty(w) = \max_{0\le k\le n} \frac{|V(w)\cos kw - U(w)\sin kw|}{\sum_{i=0}^{n}\alpha_i|\sin(k-i)w|}$$

$$y_\infty(w) = \max_{n+1\le k\le 2n+1} \frac{|U(w)\cos kw + V(w)\sin kw|}{\sum_{i=0}^{n}\alpha_i|\cos(k-i)w|},$$

then

$$r(w) = \max\{x_\infty(w),\ y_\infty(w)\}, \text{ and } r_R = \inf_{0\le w<2\pi} r(w). \tag{3.6.1}$$

Investigations of robust Schur stability have been conducted by Polyak and Tsypkin [PT3].

Example 3.6.2 l_p stability radius of Hurwitz polynomials.

Let $p+q=pq$,

$$S_p(w) = \left[|\alpha_0|^q + |\alpha_2 w^2|^q + |\alpha_4 w^4|^q \ldots\right]^{\frac{1}{q}},$$

and

$$T_p(w) = \left[|\alpha_1 w|^q + |\alpha_3 w^3|^q + |\alpha_5 w^5|^q + \ldots\right]^{\frac{1}{q}}.$$

$$x_p(w) = \frac{U(w)}{S_p(w)}, \text{ and } y_p(w) = \frac{V(w)}{T_p(w)}.$$

Then

$$r(w) = [|x_p(w)|^p + |y_p(w)|^p]^{\frac{1}{p}}, \text{ and } r_\infty = \left|\frac{a_n^0}{\alpha_n}\right|.$$

Finally we obtain

$$r_R = \min\left\{\inf_w r(w),\ \left|\frac{a_n^0}{\alpha_n}\right|\right\}. \tag{3.6.2}$$

Studies of robust stability of continuous systems has been conducted by Polyak and Tsypkin [PT1]; Tsypkin and Polyak [TP1], [TP2]; and by Robledo, Desages and Cendra [RDC].

Necessary and sufficient robust stability conditions given in Tsypkin and Polyak [TP2] are stated in an attractive graphical form. Let $\mathbf{p}^0(s) = a_0^0 + a_1^0 s + \ldots + a_n^0 s^n$ be a stable nominal polynomial, $\alpha_k > 0$, $k = 0, 1, \ldots, n$, and $1 \le p \le \infty$.

Tsypkin–Polyak criterion. The polynomial family

$$\mathbf{P}_\gamma = \left\{a_0 + a_1 s + \ldots + a_n s^n\ ;\ \left[\sum_{k=1}^{n}\left|\frac{a_k - a_k^0}{\alpha_k}\right|^p\right]^{\frac{1}{p}} \le \gamma\right\} \tag{3.6.3}$$

is stable if and only if:

1. For $w \geq 0$ the plot $z_p(w) = x_p(w) + jy_p(w)$ crosses exactly n consecutive quadrants of the complex plane in a strictly counterclockwise direction.

2. The plot does not intersect l_p circle with radius γ centered at the origin.

3. Absolute values of the coordinates of the boundary points $z_p(0)$ and $z_p(\infty)$ exceed γ.

We illustrate the criterion by a numerical example borrowed from Tsypkin and Polyak [TP2].

Example 3.6.3 Consider a Hurwitz stable nominal polynomial

$$\mathbf{p}^0(s) = 433.5 + 667.25s + 502.72s^2 + 251.25s^3 + 80.25s^4 + 14.0s^5 + 1.0s^6.$$

The measure of perturbation is given by a weighted l_∞ norm (interval constraints), and the weights α_k, $k = 0, \ldots, 6$ are given by:

α_0	α_1	α_2	α_3	α_4	α_5	α_6
43.35	33.36	25.137	15.075	5.6175	1.4	0.1

A MATLAB computation of (3.6.2) yields $r_R = 1.20$. The plot of $z_\infty(w)$ and l_∞ circle of radius 1.20 are shown below.

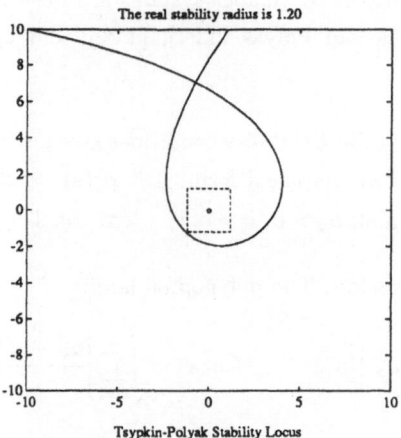

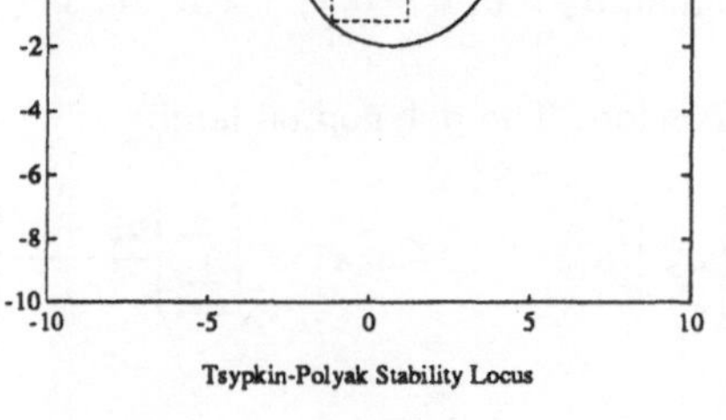

3.7. Link to Structured Matrix Perturbations

In the remaining sections we briefly relate results of the chapter to structured matrix perturbations and the structured singular value. The link between the structured singular value and stability of uncertain polynomials has been described, for example, by Chen, Fan and Nett [CFN], and Hinrichsen and Pritchard [HP3].

We start with an example due to Hinrichsen and Pritchard [HP1]. Consider the linear oscillator

$$\frac{d^2\xi}{dt^2} + a_1\frac{d\xi}{dt} + a_2\xi = 0$$

whose state space equation is

$$\frac{d}{dt}\mathbf{x} = \begin{bmatrix} 0 & 1 \\ -a_2 & -a_1 \end{bmatrix}\mathbf{x}.$$

Assume that a_1, a_2 are uncertain parameters, then

$$\frac{d}{dt}\mathbf{x} = (A + BDC)\mathbf{x}$$

where

$$A = \begin{bmatrix} 0 & 1 \\ -a_2 & -a_1 \end{bmatrix}, \; B = \begin{bmatrix} 0 \\ 1 \end{bmatrix}, \; D = [d_1 \; d_2], \; C = \begin{bmatrix} 0 & 1 \\ 1 & 0 \end{bmatrix}.$$

The matrices B and C define the structure of the perturbation. To simplify the exposition we focus on the complex stability radius whose definition is given next.

Definition 3.7.1 *The complex stability radius of $A \in \mathbf{C}^{n\times n}$ with respect to the matrices $B \in \mathbf{C}^{n\times m}$, $C \in \mathbf{C}^{k\times n}$, and the open region Ω in the complex plane is defined by*

$$r_C(A; B, C, \Omega) = \inf\{||D||, \; \sigma(A + BDC) \cap \Omega^c \neq \emptyset\}.$$

The expression $||D||$ denotes the operator norm of the matrix D and will be defined in the next section, σ stands for the spectrum. The investigation of the complex stability radius given in this chapter follows ideas of Hinrichsen and Pritchard [HP2]. We show that the complex stability radius for polynomial families can be obtain as a special particular case of the matrix perturbation problem. Next we consider multiple matrix perturbations, and discuss the link to the structured singular value introduced by Doyle in [D]. For further details concerning matrix perturbations we refer to [HP2] and references therein. A specific treatment of the real stability radius is provided in [QBRDYD].

3.8. Matrix Perturbations

We shall say that a matrix D is a destabilizing perturbation if there exists $s \in \Omega^c$ and a nonzero $\mathbf{x} \in \mathbf{C}^n$ such that

$$(A + BDC)\mathbf{x} = s\mathbf{x}.$$

The transfer matrix $G(s) = C(sI - A)^{-1}B$ will play an important role in the characterization of destabilizing perturbations given next.

Proposition 3.8.1 An $m \times k$ matrix D is destabilizing if and only if there exists $s \in \Omega^c$ and a nonzero $\mathbf{y} \in \mathbf{C}^k$ such that

$$[I - G(s)D]\,\mathbf{y} = 0. \tag{3.8.1}$$

Proof: If $\sigma(A + BDC) \cap \Omega^c \neq \emptyset$, then there exists a nonzero $\mathbf{x} \in \mathbf{C}^n$, and $s \in \Omega^c$ such that

$$(A + BDC)\mathbf{x} = s\mathbf{x}, \text{ and } BDC\mathbf{x} = [sI - A]\,\mathbf{x}.$$

This implies

$$[sI - A]^{-1}\, BDC\mathbf{x} = \mathbf{x},\ C\mathbf{x} \neq 0, \text{ and } C\,[sI - A]^{-1}\, BDC\mathbf{x} = C\mathbf{x}.$$

For $\mathbf{y} = C\mathbf{x}$ we obtain (3.8.1). Conversely, if (3.8.1) holds we set $\mathbf{x} = [sI - A]^{-1}\, BD\mathbf{y}$ and obtain $s\mathbf{x} = A\mathbf{x} + BD\mathbf{y}$. Since $C\mathbf{x} = G(s)D\mathbf{y} = \mathbf{y}$ one has $s\mathbf{x} = A\mathbf{x} + BDC\mathbf{x}$, i.e., D is destabilizing. □

Let $||\cdot; m||$ and $||\cdot; k||$ be norms in complex spaces $\mathbf{C}^m$, and $\mathbf{C}^k$ respectively. For an $m \times k$ matrix D we denote the corresponding operator norms by

$$\begin{aligned} ||m, D; k|| &= \sup_{\mathbf{x} \neq 0} \left\{ \frac{||\mathbf{x}D; k||}{||\mathbf{x}; m||} \ :\ \mathbf{x} = (x_1, \ldots, x_m) \in C^m \right\} \\ ||m; D, k|| &= \sup_{\mathbf{y} \neq 0} \left\{ \frac{||D\mathbf{y}; m||}{||\mathbf{y}; k||} \ :\ \mathbf{y} = (y_1, \ldots, y_k)^t \in C^k \right\}. \end{aligned} \tag{3.8.2}$$

For example, if $k = 1$, $||a + jb; 1|| = |a + jb| = \sqrt{a^2 + b^2}$, $||\mathbf{z}; m|| = \left[\sum_{i=0}^{m-1} \left| \frac{z_i}{\alpha_i} \right|^q \right]^{1/q}$, and D is a column vector $(D_0, \ldots, D_{m-1})^t$, then due to Hölder's inequality with $p + q = pq$ one has

$$||m, D; 1|| = \left[\sum_{i=0}^{m-1} |\alpha_i D_i|^p \right]^{1/p}, \quad \text{and} \quad ||m; D, 1|| = \left[\sum_{i=0}^{m-1} \left| \frac{D_i}{\alpha_i} \right|^q \right]^{1/q}. \tag{3.8.3}$$

In particular (3.8.1) implies

$$||m; D, k|| \cdot ||k; G(s), m|| \geq 1. \tag{3.8.4}$$

If $r_C(A; B, C, \Omega) < \infty$ an easy compactness argument shows the existence of a minimum norm destabilizing perturbation D with $||m; D, k|| = r_C(A; B, C, \Omega)$.

To derive a formula for the complex stability radius we next prove a modification of Gastiner–Kahan Theorem (see e.g., Kahan [Kah], or Martin [M]).

Theorem 3.8.1 Let $\mathcal{A}$ be a nonsingular $n \times n$ matrix. Given an $n \times m$ matrix B, an $m \times k$ matrix D, and a $k \times n$ matrix C

$$\min\{||m; D, k|| \;,\; \mathcal{A} + BDC \text{ is singular}\} = \begin{cases} \infty & \text{if} \quad ||k; C\mathcal{A}^{-1}B, m|| = 0. \\ [||k; C\mathcal{A}^{-1}B, m||]^{-1} & \text{otherwise.} \end{cases}$$

Proof: First consider the case when $||k; C\mathcal{A}^{-1}B, m|| = 0$. We show that in this case the matrix $\mathcal{A} + BDC$ is always nonsingular. Indeed suppose the opposite, i.e., there exist a matrix D, and $\mathbf{x} \neq \mathbf{0}$ such that $(\mathcal{A} + BDC)\mathbf{x} = \mathbf{0}$, that is $\mathcal{A}\mathbf{x} = -(BDC)\mathbf{x}$, and $\mathbf{0} \neq \mathbf{x} = -\mathcal{A}^{-1}(BDC)\mathbf{x}$. This in turn implies that $\mathbf{0} \neq C\mathbf{x}$, and

$$\mathbf{0} \neq C\mathbf{x} = -[C\mathcal{A}^{-1}B]DC\mathbf{x} = \mathbf{0}.$$

From now on we assume that $||k; C\mathcal{A}^{-1}B, m|| \neq 0$. Suppose first that $\mathcal{A} + BDC$ is singular. There exists $\mathbf{x} \neq \mathbf{0}$ such that $(\mathcal{A} + BDC)\mathbf{x} = \mathbf{0}$, $\mathcal{A}\mathbf{x} = -(BDC)\mathbf{x}$, $\mathbf{x} = -\mathcal{A}^{-1}(BDC)\mathbf{x}$, and $C\mathbf{x} = -C\mathcal{A}^{-1}(BDC)\mathbf{x}$. If $\mathbf{y} = C\mathbf{x}$, then $\mathbf{y} \neq \mathbf{0}$, and $\mathbf{y} = -[C\mathcal{A}^{-1}B]D\mathbf{y}$. That is

$$1 \leq ||k; [C\mathcal{A}^{-1}B]D, k|| \leq ||k; C\mathcal{A}^{-1}B, m|| \cdot ||m; D, k||,$$

and

$$\frac{1}{||k; C\mathcal{A}^{-1}B, m||} \leq ||m; D, k||. \tag{3.8.5}$$

Let now $\mathbf{y}$ be a vector in $\mathbf{C}^m$ such that

$$1 = ||C\mathcal{A}^{-1}B\mathbf{y}; k|| = ||k; C\mathcal{A}^{-1}B, m|| \cdot ||\mathbf{y}; m||,$$

i.e.,

$$||\mathbf{y}; m|| = \frac{1}{||k; C\mathcal{A}^{-1}B, m||}. \tag{3.8.6}$$

Due to the Hahn–Banach Theorem there exists a linear functional Λ such that

$$||1; \Lambda, k|| = 1, \text{ and } \Lambda(C\mathcal{A}^{-1}B\mathbf{y}) = 1. \tag{3.8.7}$$

Let $D = -\mathbf{y}\Lambda$, then

$$(\mathcal{A} + BDC)\mathcal{A}^{-1}B\mathbf{y} = B\mathbf{y} + B[-\mathbf{y}\Lambda C\mathcal{A}^{-1}B\mathbf{y}] = B\mathbf{y} - B\mathbf{y} = \mathbf{0}.$$

So $\mathcal{A} + BDC$ is singular. On the other hand according to (3.8.6), and (3.8.7)

$$||m; D, k|| = ||m; \mathbf{y}\Lambda, k|| \leq ||\mathbf{y}; m|| \cdot ||1; \Lambda, k|| \leq \frac{||1; \Lambda, k||}{||k; C\mathcal{A}^{-1}B, m||} = \frac{1}{||k; C\mathcal{A}^{-1}B, m||}.$$

This inequality along with (3.8.5) completes the proof of the theorem. □

Theorem 3.8.2 Let A be an Ω stable $n \times n$ complex matrix, and

$$G(s) = C\,(sI - A)^{-1}\,B,$$

then

$$r_C(A; B, C, \Omega) = \begin{cases} \infty & \text{if } \max_w ||k; G(\delta(w)), m|| = 0 \\ \left[\max_w ||k; G(\delta(w)), m||\right]^{-1} & \text{otherwise.} \end{cases} \tag{3.8.8}$$

Proof: Let D be a matrix such that $A + BDC$ is unstable, i.e., there exists $s \in \sigma(A+BDC) \cap \Omega^c$. Since $\sigma(A) \subset \Omega$ the matrix $sI - A$ is nonsingular. If $s = \delta(w) \in \partial\Omega$, then we apply Theorem 3.8.1 to the matrices $\mathcal{A} = \delta(w)I - A$, B, C, and obtain

$$||m; D, k|| \geq [||k; G(\delta(w)), m||]^{-1}. \tag{3.8.9}$$

If $s \in \text{Int}\ \Omega^c$, then there exists $0 < \lambda < 1$ such that $\sigma(A + B[\lambda D]C) \cap \partial\Omega \neq \emptyset$. Hence the exists w with $\delta(w) \in \sigma(A + B[\lambda D]C)$, and

$$||m; D, k|| > ||m; \lambda D, k|| \geq [||k; G(\delta(w)), m||]^{-1}. \tag{3.8.10}$$

This shows that

$$r_C(A; B, C, \Omega) \geq \left[\max_w ||k; G(\delta(w)), m||\right]^{-1}. \tag{3.8.11}$$

If $\max_w ||k; G(\delta(w)), m|| = 0$, then $r_C(A; B, C, \Omega) = \infty$, and the proof is completed. Otherwise let $||k; G(\delta(w_0)), m|| = \max_w ||k; G(\delta(w)), m|| \neq 0$, then there exist $\mathbf{y} \in \mathbf{C}^m$, and a linear functional Λ such that

$$||G(\delta(w_0))\mathbf{y}; k|| = ||k; G(\delta(w_0)), m|| \cdot ||\mathbf{y}; m||,\ \ ||1; \Lambda, k|| = 1,\ \text{ and } \Lambda G(\delta(w_0))\mathbf{y} = 1.$$

Then, according to (3.8.11) for $D = -\mathbf{y}\Lambda$ one has $||k; D, m|| \leq ||k; G(\delta(w_0)), m||$. This completes the proof. □

We now show that the formula for the complex stability radius given in Theorem 3.1.3 is a special case of (3.8.8). Consider two $(n+1)\times(n+1)$ matrices I_n, P, and an $(n+1)\times 1$ matrix B.

$$I_n = \begin{bmatrix} 1 & 0 & 0 & \dots & 0 \\ 0 & 1 & 0 & \dots & 0 \\ & & & & \\ 0 & 0 & \dots & 1 & 0 \\ 0 & 0 & 0 & \dots & 0 \end{bmatrix}, \; P = \begin{bmatrix} 0 & 1 & 0 & 0 & \dots & 0 \\ 0 & 0 & 1 & 0 & \dots & 0 \\ & & & & & \\ 0 & 0 & 0 & 0 & \dots & 1 \\ -t_0^0 & -t_1^0 & -t_2^0 & -t_3^0 & \dots & -t_n^0 \end{bmatrix}, \text{ and } B = \begin{bmatrix} 0 \\ 0 \\ \\ 0 \\ 1 \end{bmatrix}.$$

Then

$$\det\left[sI_n - P\right] = t_0^0 + t_1^0 s + t_2^0 s^2 + \ldots + t_n^0 s^n = \mathbf{p}(s, \mathbf{0}).$$

So the nominal polynomial $\mathbf{p}(s,\mathbf{0})$ is Ω stable if and only if the matrix $sI_n - P$ is nonsingular for each $s \in \Omega^c$. Furthermore,

$$\det\left[sI_n - (P - B\mathbf{d}T)\right] = t_0(\mathbf{d}) + t_1(\mathbf{d})s + \ldots + t_n(\mathbf{d})s^n = \mathbf{p}(s, \mathbf{d}),$$

and $\mathbf{p}(\delta(w),\mathbf{d}) = 0$ if and only if the matrix $[\delta(w)I_n - P] + B\mathbf{d}T$ is singular. The matrix $T\left[sI_n - P\right]^{-1}B$ is an m dimensional column vector with the norm given by $\left\|1, T\left[sI_n - P\right]^{-1}B; m\right\|$. Theorem 3.8.2 provides the formula for $r_C(P;B,T,\Omega)$ as follows

$$r_C(P;B,T,\Omega) = \begin{cases} +\infty & \text{if } \left\|1, T\left[\delta(w)I_n - P\right]^{-1}B; m\right\| \equiv 0, \\ \\ \min\limits_{w}\left[\left\|1, T\left[\delta(w)I_n - P\right]^{-1}B; m\right\|\right]^{-1} & \text{otherwise.} \end{cases}$$

We note that

$$\left[sI_n - P\right]^{-1}B = \frac{1}{\mathbf{p}(s,\mathbf{0})}\begin{bmatrix} 1 \\ s \\ s^2 \\ \\ s^n \end{bmatrix} = \frac{\mathbf{s}}{\mathbf{p}(s,\mathbf{0})}, \text{ where } \mathbf{s} = \begin{bmatrix} 1 \\ s \\ s^2 \\ \\ s^n \end{bmatrix}.$$

Hence

$$T\left[\delta(w)I_n - P\right]^{-1}B = T\left[\frac{\boldsymbol{\delta}(w)}{\mathbf{p}(\delta(w),\mathbf{0})}\right], \text{ where } \boldsymbol{\delta}(w) = \begin{bmatrix} 1 \\ \delta(w) \\ \delta(w)^2 \\ \\ \delta(w)^n \end{bmatrix}.$$

In other words

$$r_C(P;B,T,\Omega) = \begin{cases} +\infty & \text{if } \left\| 1, T\left[\frac{\delta(w)}{p(\delta(w),0)}\right]; m\right\| \equiv 0, \\ \min_w \left[\left\| 1, T\left[\frac{\delta(w)}{p(\delta(w),0)}\right]; m\right\|\right]^{-1} & \text{otherwise.} \end{cases}$$

This shows that $r_C(P;B,T,\Omega) = r_C$ given in Theorem 3.3.1

3.9. Multiple Matrix Perturbations and μ

Not all parameter uncertainties can be represented by a single perturbation structure $A + BDC$ (see e.g., [HP2]) . In this section we consider perturbations

$$A + \sum_{i=1}^{N} B_i \Delta_i C_i \text{ where } B_i \in \mathbb{C}^{n \times m_i},\ \Delta_i \in \mathbb{C}^{m_i \times k_i},\ C_i \in \mathbb{C}^{k_i \times n},\ i = 1, \ldots, N. \tag{3.9.1}$$

If

$$D = \begin{bmatrix} \Delta_1 & & 0 \\ & \cdot & \\ 0 & & \Delta_N \end{bmatrix}, \tag{3.9.2}$$

then the size of the perturbation D is measured by

$$||D|| = \max ||m_i; \Delta_i, k_i|| \,.$$

Definition 3.9.1 *The complex stability radius of A with respect to the perturbation structure (3.9.1) is defined by*

$$r_C\left(A; [B_i, C_i, N], \Omega\right) = \inf\left\{ ||D||,\ \sigma\left(A + \sum_{i=1}^{N} B_i \Delta_i C_i\right) \cap \Omega^c \neq \emptyset \right\}.$$

Just as in the single perturbation case an important role is played by the block transfer matrix $G(s) \in \mathbb{C}^{k \times m}$, $k = \sum_{i=1}^{N} k_i$, $m = \sum_{i=1}^{N} m_i$,

$$G(s) = \begin{bmatrix} G_{11}(s) & \cdots & G_{1N}(s) \\ \cdots & \cdots & \cdots \\ \cdots & \cdots & \cdots \\ G_{N1}(s) & \cdots & G_{NN}(s) \end{bmatrix}, \qquad \begin{matrix} G_{ij}(s) = C_i(sI - A)^{-1} B_j, \\ \\ i, j = 1, \ldots, N. \end{matrix} \tag{3.9.3}$$

The following characterization of destabilizing perturbations holds for multiple matrix perturbations.

Proposition 3.9.1 A perturbation matrix $D = \text{diag}\ (\Delta_1, \ldots, \Delta_N)$ is destabilizing if and only if there exists $s \in \Omega^c$ and a nonzero $\mathbf{y} = (\mathbf{y}_1, \ldots, \mathbf{y}_N)^t \in \mathbf{C}^k$, $\mathbf{y}_i \in \mathbf{C}^{k_i}$ such that

$$[I - G(s)D]\,\mathbf{y} = 0, \text{ and } \|D\| \cdot \|G(s)\| \geq 1. \tag{3.9.4}$$

Furthermore

$$r_C\,(A; [B_i, C_i, N], \Omega) \geq \left[\max_{w \in I_\Omega} \|G(\delta(w))\|\right]^{-1}.$$

The proof of the proposition is identical to that of Proposition 3.8.1 and will be omitted.

In [D] Doyle introduced the concept of structured singular values.

Definition 3.9.2 *The structured singular value $\mu(M)$ of a matrix $M \in \mathbf{C}^{k \times m}$ with respect to the structure given by (3.9.2) is*

$$\mu(M) = \left[\min\{\|D\|,\ \ det\ (I - MD) = 0\}\right]^{-1}. \tag{3.9.5}$$

As an immediate corollary to Proposition 3.9.1 we have

$$r_C\,(A; [B_i, C_i, N], \Omega) = \left[\sup_{w \in I_\Omega} \mu(G(\delta(w)))\right]^{-1}. \tag{3.9.6}$$

The structured singular value is a general linear algebra tool that provides necessary and sufficient conditions for structured matrix perturbation problems. Evaluations of the structured singular value and its computable bounds is an important, difficult and challenging problem. For further details concerning μ the reader may consult, for example, [PD], [PDB] and references therein.

A generalization of the structured singular value μ has been introduced recently by Barmish and Polyak [BP]. The *volumetric singular value* includes no assumption that all the components Δ_i of the uncertainty matrix D are expanded by the same factor. The volumetric singular value allows one to systematically study the tradeoffs associated with various uncertainty components. At the same time $\mu_v(M)$ enjoys many of the nice computational properties enjoyed by $\mu(M)$.

Chapter 4

Multiaffine Polynomial Families

Zero exclusion criterion plays an important role in robustness analysis. When the value set $\mathcal{P}_w$ is a convex polygon in the complex plane the condition $0 \notin \mathcal{P}_w$ is easy to check. When one is concerned with the system $\dot{\mathbf{x}} = A\mathbf{x}$ where the entries of the matrix A are allowed to vary in corresponding intervals the characteristic polynomials of the interval linear system are

$$\mathbf{P} = \{\mathbf{p}(s, \mathbf{x}) = a_0(\mathbf{x}) + a_1(\mathbf{x})s + \ldots + a_{n-1}(\mathbf{x})s^{n-1} + a_n(\mathbf{x})s^n, \ \mathbf{x} \in \mathbf{B} \subseteq \mathbf{R}^m\}, \quad (4.0.1)$$

where $a_i(\mathbf{x})$, $i = 0, 1, \ldots, n-1$ are multiaffine mappings from $\mathbf{R}^m$ to $\mathbf{R}$, and

$$\mathbf{B} = \{\mathbf{x} \ : \ \mathbf{x} \in \mathbf{R}^m, \ \underline{x}_k \le x_k \le \overline{x}_k, \ \overline{x}_k - \underline{x}_k > 0, \ k = 1, \ldots, m\}$$

is a box in $\mathbf{R}^m$. Since the linear mapping

$$(x_1, \ldots, x_m)^t \to \left(\frac{x_1 - \underline{x}_1}{\overline{x}_1 - \underline{x}_1}, \ldots, \frac{x_m - \underline{x}_m}{\overline{x}_m - \underline{x}_m}\right)^t$$

preserves multiaffine properties in order to simplify the exposition we also consider the unit box

$$\mathbf{B}^m = \{\mathbf{x} \ : \ \mathbf{x} \in \mathbf{R}^m, \ 0 \le x_k \le 1, \ k = 1, \ldots, m\},$$

for convenience. Testing families for polynomials with general multiaffine dependence on uncertain parameters generically do not exist (see e.g., Ackermann [A], and Holohan and Safonov [HS]). Additional technical restrictions are needed if one wants to generate "easily testable" criteria for multiaffine polynomial families.

The value sets of the multiaffine polynomial family (4.0.1) are given by

$$\mathcal{P}_w = \left\{a_0(\mathbf{x}) + a_1(\mathbf{x})\delta(w) + \ldots + a_{n-1}(\mathbf{x})[\delta(w)]^{n-1} + a_n(\mathbf{x})[\delta(w)]^n, \ \mathbf{x} \in \mathbf{B}^m\right\}.$$

For a fixed w the mapping $\mathbf{x} \to \mathbf{p}(\delta(w), \mathbf{x})$ is a multiaffine mapping f from $\mathbf{B}^m$ to the complex plane $\mathbf{C}$, i.e.,

$$f(x_1, \ldots, x_k + \gamma, \ldots, x_m) - f(x_1, \ldots, x_m) = \gamma \frac{\partial f}{\partial x_k}(\mathbf{x}) = \gamma G_k(\mathbf{x}), \; k = 1, \ldots, m.$$

It is of interest to know when the value sets $\mathcal{P}_w$ and/or their boundaries $\partial \mathcal{P}_w$ can be easily described, and simple methods can be devised to determine if they contain the origin. The Mapping Theorem [ZD] implies $f(\mathbf{B}^m) \subseteq \operatorname{conv} f(\mathbf{V}^m)$. The following conjecture has been made by Hollot and Xu [HX]:

Conjecture 4.0.1 $f(\mathbf{B}^m)$ *is a polygon if and only if all the edges of* $\operatorname{conv} f(\mathbf{V}^m)$ *are the images of edges of* $\mathbf{B}^m$.

The conjecture has been resolved independently in different forms by Anderson, Kraus, Mansour and Dasgupta [AKMD], Polyak [Po], and Tsing and Tits [TT]. It turns out that even when $\partial \operatorname{conv} f(\mathbf{V}^m) \subseteq f(\mathbf{B}^m)$, the set $f(\mathbf{B}^m)$ may fail to be simply connected (see [AKMD]).

A special class of *orientation preserving multiaffine transformations* have been introduced in [AKMD] and [Po]. The orientation preserving multiaffine transformations f satisfy

$$f(\mathbf{B}^m) = \operatorname{conv} f(\mathbf{V}^m). \tag{4.0.2}$$

In the next section we reproduce the results of [AKMD] and [Po] and show that the image of a box under an orientation preserving multiaffine transformation is a convex polygon in the complex plane. Furthermore, our proof shows that the image of the two dimensional faces of the box covers the polygon.

Although mathematically elegant, in many simple cases of practical interest these conditions fail (see e.g., [A]). On the other hand, as we show by an example, the Mapping Theorem may be excessively conservative.

Example 4.0.1 Consider the polynomial family

$$\mathbf{P} = \left\{ \mathbf{p}(s, \mathbf{x}) = \prod_{k=1}^{m} (1 + x_k s), \; \underline{x_k} \le x_k \le \overline{x_k} \right\}.$$

It is easy to see that the family $\mathbf{P}$ is Schur stable if $1 < \underline{x_k}$, $k = 1, \ldots, m$. To simplify the presentation we assume that the intervals $\left[\underline{x_k}, \overline{x_k}\right]$ are identical, i.e.,

$$1 < \underline{x} = \underline{x_k}, \text{ and } \underline{x} < \overline{x} = \overline{x_k}, \; k = 1, \ldots, m.$$

Note that the images of the vertices of the box under the mapping $\mathbf{x} \rightarrow \mathbf{p}(e^{jw}, \mathbf{x})$ are given by

$$\left(1 + e^{jw\underline{x}}\right)^m \left(\frac{1 + e^{jw\overline{x}}}{1 + e^{jw\underline{x}}}\right)^i, \quad i = 0, \ldots, m. \tag{4.0.3}$$

When $\arg \dfrac{1 + e^{jw\overline{x}}}{1 + e^{jw\underline{x}}} > \dfrac{\pi}{m}$, the $(m+1)$ points (4.0.3) encircle the origin, and the Mapping Theorem is inconclusive.

In the subsequent sections we follow the ideas of [PK] and describe the boundary of $f(\mathbf{B}^m)$ for a general multiaffine transformations of $\mathbf{R}^m$. The general criterion obtained in [PK] provides necessary and sufficient stability conditions for polynomial families whose coefficients are multiaffine functions of parameters. The criterion is applied to two particular cases: systems with a cascade of first order uncertain blocks, and closed loop systems with interval plant and interval controller . A numerical example completes the chapter.

4.1. Orientation Preserving Multiaffine Transformation

In this section we describe a special class of multiaffine transformations that map an m dimensional box into a convex polygon.

Definition 4.1.1 *Let* $\mathbf{X}$ *be a subset of* $\mathbf{R}^m$. *A multiaffine function* f *is an* $\mathcal{O}_{\mathbf{X}}$ *(orientation preserving in* $\mathbf{X}$*) function if* $\forall \mathbf{x} \in \mathbf{X}$, *and for each integer* k *one has*

$$G_k(\mathbf{x}) \prec G_{k+i}(\mathbf{x}), \quad i = 1, \ldots, m-1, \tag{4.1.1}$$

where $G_k(\mathbf{x}) = \dfrac{\partial f}{\partial x_k}(\mathbf{x})$, $G_{k+2m}(\mathbf{x}) = G_k(\mathbf{x})$, *and* $G_{k+m}(\mathbf{x}) = -G_k(\mathbf{x})$, $k = 1, \ldots, m$.

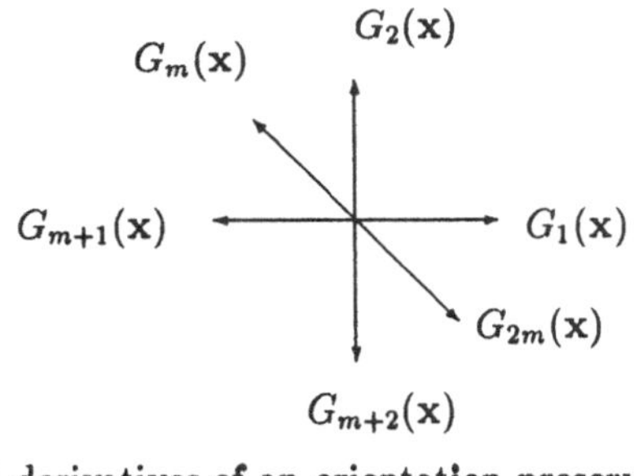

Partial derivatives of an orientation preserving function f

Theorem 4.1.1 (Anderson, Kraus, Mansour, Dasgupta [AKMD]). Let f be a multiaffine $\mathcal{O}_{\mathbf{V}^m}$-function. Then

- The $2m$ vertices

Name	Vertex
$\mathbf{v}_p^1$	$(0,0,\ldots,0,0)$
$\mathbf{v}_p^2$	$(1,0,\ldots,0,0)$
...	...
$\mathbf{v}_p^m$	$(1,1,\ldots,1,0)$
$\mathbf{v}_p^{m+1}$	$(1,1,\ldots,1,1)$
$\mathbf{v}_p^{m+2}$	$(0,1,\ldots,1,1)$
...	...
$\mathbf{v}_p^{2m}$	$(0,0,\ldots,0,1)$

  are the principal vertices of the box $\mathbf{B}^m$.

- $f(\mathbf{B}^m)$ is a convex polygon conv $\{f(\mathbf{v}_p^1),\ldots,f(\mathbf{v}_p^{2m})\}$ whose consecutive vertices are $f(\mathbf{v}_p^1),\ldots,f(\mathbf{v}_p^{2m})$.

Remark 4.1.1 A vertex $\mathbf{v} = (v_1,\ldots,v_m)^t$ is a principal vertex if and only if there exists a half plane that contains the complex numbers

$$(-1)^{v_1}G_1(\mathbf{v}),\ldots,(-1)^{v_m}G_m(\mathbf{v}).$$

A half plane in the complex plane is defined by a non zero complex number G. The vertex $\mathbf{v}$ is a principal vertex if and only if there exists a non zero complex number $G = G(\mathbf{v})$ such that

$$(-1)^{v_i}G_i(\mathbf{v}) \preceq G \text{ for each } i = 1,\ldots,m. \tag{4.1.2}$$

The proof of the theorem given below combines arguments given in [AKMD] and [Po]. We first consider the two dimensional box $\mathbf{B}^2$.

4.1.1. Image of a Two Dimensional Box

The four principal vertices of the box $\mathbf{B}^2$ are

Name	Vertex
$\mathbf{v}_p^1$	$(0,0)$
$\mathbf{v}_p^2$	$(1,0)$
$\mathbf{v}_p^3$	$(1,1)$
$\mathbf{v}_p^4$	$(0,1)$

Furthermore, the following result holds in the two parameter case.

Lemma 4.1.1 *A multiaffine $\mathcal{O}_{\mathbf{V}^2}$ function f is an $\mathcal{O}_{\mathbf{B}^2}$ function.*

Proof: The relations

$$G_1(0,x_2) \prec G_2(0,x_2), \text{ and } G_1(1,x_2) \prec G_2(1,x_2), \tag{4.1.3}$$

imply

$$G_1(\lambda,x_2) \prec G_2(\lambda,x_2), \text{ for each } 0 \le \lambda \le 1. \tag{4.1.4}$$

Indeed, $G_1(\lambda, x_2)$ does not depend on λ, and $G_2(\lambda, x_2)$ is a linear function of λ. The boundary conditions (4.1.3) enforce (4.1.4). Analogously one has

$$G_1(x_1,\lambda) \prec G_2(x_1,\lambda), \text{ for each } 0 \le \lambda \le 1. \tag{4.1.5}$$

A combination of (4.1.4) and (4.1.5) completes the proof.

Lemma 4.1.2 *Suppose that $\mathbf{v} \neq \mathbf{v}_p^i$, and $\mathbf{v} \neq \mathbf{v}_p^{i+1}$. Then*

$$f(\mathbf{v}_p^{i+1}) - f(\mathbf{v}_p^i) \prec f(\mathbf{v}) - f(\mathbf{v}_p^i). \tag{4.1.6}$$

Proof: To simplify the exposition we assume that $i = 1$. Then $\mathbf{v} = \mathbf{v}_p^3$, or $\mathbf{v} = \mathbf{v}_p^4$. If $\mathbf{v} = \mathbf{v}_p^3$, then one has

$$\begin{aligned} f(\mathbf{v}_p^2) - f(\mathbf{v}_p^1) = G_1(\mathbf{v}_p^1) \prec G_2(\mathbf{v}_p^2) + G_1(\mathbf{v}_p^1) &= f(\mathbf{v}_p^3) - f(\mathbf{v}_p^2) + f(\mathbf{v}_p^2) - f(\mathbf{v}_p^1) \\ &= f(\mathbf{v}_p^3) - f(\mathbf{v}_p^1). \end{aligned}$$

If $\mathbf{v} = \mathbf{v}_p^4$ the result follows from condition (4.1.1).

In particular Lemma 4.1.2 shows that the boundary of conv $f(\mathbf{B}^2)$ is covered by the images of four edges of $\mathbf{B}^2$, i.e.,

$$\partial \text{conv } f(\mathbf{B}^2) \subseteq \bigcup_{\substack{0 \le \lambda \le 1 \\ i = 1,2,3,4}} f\left(\lambda \mathbf{v}_p^{i+1} + (1-\lambda)\mathbf{v}_p^i\right).$$

We next show that conv $f(\mathbf{B}^2) = f(\mathbf{B}^2)$.

Definition 4.1.2 *For a set Z in the complex plane, and a complex number z_0 the distance $d(z_0, Z)$ is defined by*

$$d(z_0, Z) = \inf\left\{|z_0 - z| \ : \ z \in Z\right\}.$$

Lemma 4.1.3 *For each $z \in$ conv $f(\mathbf{B}^2)$ there exists $\mathbf{x} \in \mathbf{B}^2$ so that $z = f(\mathbf{x})$.*

Proof: Suppose the opposite, i.e., there exists $z_0 \in \text{conv } f(\mathbf{B}^2)$, with $d\left(z_0, f(\mathbf{B}^2)\right) > 0$. Let $z = f(\mathbf{x})$ be such that

$$|z_0 - z| = d\left(z_0, f(\mathbf{B}^2)\right). \tag{4.1.7}$$

A simple convexity argument along with Lemma 4.1.2 show that $\mathbf{x}$ does not belong to the edges of $\mathbf{B}^2$, and $\mathbf{x}$ is an interior point of $\mathbf{B}^2$. The condition $G_1(\mathbf{x}) \prec G_2(\mathbf{x})$ immediately leads to a contradiction.

4.1.2. Image of a Box Under Orientation Preserving Transformation

The proof of the general case will proceed by induction. The basic step of the induction is $m = 2$. For $\mathbf{x} \in \mathbf{B}^m$ define three sets of indices $\underline{I}(\mathbf{x})$, $I_f(\mathbf{x})$, and $\overline{I}(\mathbf{x})$ as follows

$$\underline{I}(\mathbf{x}) = \{i \ : \ 0 = x_i\}, \ I_f(\mathbf{x}) = \{i \ : \ 0 < x_i < 1\}, \ \overline{I}(\mathbf{x}) = \{i \ : \ x_i = 1\}.$$

Definition 4.1.3 *For $\mathbf{x} \in \mathbf{B}^m$, x_i is a free coordinate if $i \in I_f$. Otherwise x_i is an extremal coordinate of $\mathbf{x}$.*

Lemma 4.1.4 *Suppose that $\mathbf{v} \neq \mathbf{v}_p^i$, and $\mathbf{v} \neq \mathbf{v}_p^{i+1}$. Then*

$$f(\mathbf{v}_p^{i+1}) - f(\mathbf{v}_p^i) \prec f(\mathbf{v}) - f(\mathbf{v}_p^i). \tag{4.1.8}$$

Proof: To simplify the exposition we assume that $i = 1$. An $m-1$ dimensional face of $\mathbf{B}^m$ is the set

$$\mathbf{F}_k(e) = \left\{(x_1, \ldots, x_{k-1}, e, x_{k+1}, \ldots, x_m)^t\right\}, \quad \text{where } e = 0, \text{ or } e = 1.$$

The face $\mathbf{F}_k(e)$ contains $\mathbf{v}_p^1$ and $\mathbf{v}_p^2$ if and only if $k \geq 2$, and $e = 0$. If $\mathbf{v}_p^1$ and $\mathbf{v}_p^2$ belong to the face $\mathbf{F}_k(e)$, then, due to induction, $f(\mathbf{v}_p^1)$ and $f(\mathbf{v}_p^2)$ are consecutive vertices of the convex polygon $f(\mathbf{F}_k(e))$. If $\mathbf{v} \in \mathbf{V}^m$, with $v_k = 0$ for some $k \geq 0$, then $\mathbf{v} \in \mathbf{F}_k(0)$, and (4.1.8) follows by induction. To complete the proof we have to show that the remaining two vertices

$$(0, 1, 1, \ldots, 1)^t, \text{ and } (1, 1, 1, \ldots, 1)^t$$

also satisfy (4.1.8). The corresponding proofs are identical, and we prove (4.1.8) for for the first vertex only. To check the relation for $\mathbf{v} = (0, 1, 1, \ldots, 1)^t$ consider the $m-1$ dimensional face $\mathbf{F}_1(0)$. The vertices $\mathbf{v}_p^{2m}$, $\mathbf{v}_p^1$, and $\mathbf{w} = (0, 1, 0, \ldots, 0)^t$ belong to $\mathbf{F}_1(0)$. The complex numbers $f(\mathbf{v}_p^{2m})$, $f(\mathbf{v}_p^1)$, and $f(\mathbf{w})$ are consecutive vertices of the convex polygon $f(\mathbf{F}_1(0))$. The induction assumptions imply

$$f(\mathbf{v}_p^1) - f(\mathbf{v}_p^{2m}) \prec f(\mathbf{v}) - f(\mathbf{v}_p^{2m}), \text{ and } f(\mathbf{w}) - f(\mathbf{v}_p^1) \prec f(\mathbf{v}) - f(\mathbf{v}_p^1).$$

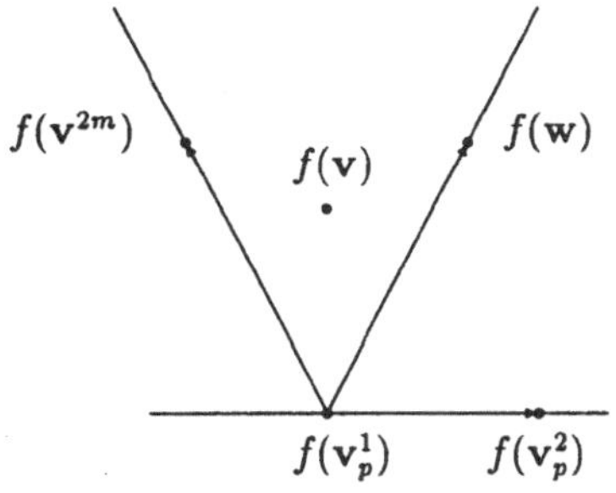

This shows that $f(\mathbf{v})$ belongs to the cone generated by $f(\mathbf{v}^{2m}) - f(\mathbf{v}_p^1)$, $f(\mathbf{w}) - f(\mathbf{v}_p^1)$, and $f(\mathbf{v}_p^1)$. This completes the proof of the lemma.

To complete the proof of the theorem we have to show that

$$\operatorname{conv} f(\mathbf{B}^m) \subseteq f(\mathbf{B}^m). \tag{4.1.9}$$

To this end the following simple property of f will be useful.

Lemma 4.1.5 *Let* $\mathbf{v}$ *be a vertex of* $\mathbf{B}^m$, *and*

$$\mathbf{v} \notin \left\{\mathbf{v}_p^1, \dots, \mathbf{v}_p^{2m}\right\}. \tag{4.1.10}$$

There exists a small square centered at $f(\mathbf{v})$ *and covered by* $f(\mathbf{B}^m)$.

Proof: Let $D_i(\mathbf{x})$, $i = 1, 2, \dots, 2m$ be complex numbers defined by

$$\begin{array}{llll} D_i(\mathbf{x}) = G_i(\mathbf{x}) & \text{and} & D_{i+m}(\mathbf{x}) = 0 & \text{if } i \in \underline{I}(\mathbf{x}) \\ D_i(\mathbf{x}) = G_i(\mathbf{x}) & \text{and} & D_{i+m}(\mathbf{x}) = -G_i(\mathbf{x}) & \text{if } i \in I_f(\mathbf{x}) \\ D_i(\mathbf{x}) = 0 & \text{and} & D_{i+m}(\mathbf{x}) = -G_i(\mathbf{x}) & \text{if } i \in \overline{I}(\mathbf{x}) \end{array} \tag{4.1.11}$$

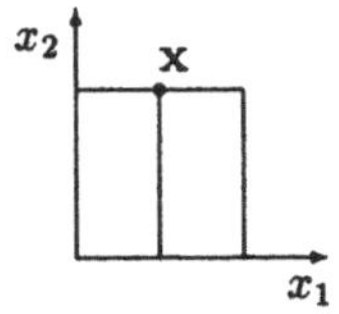

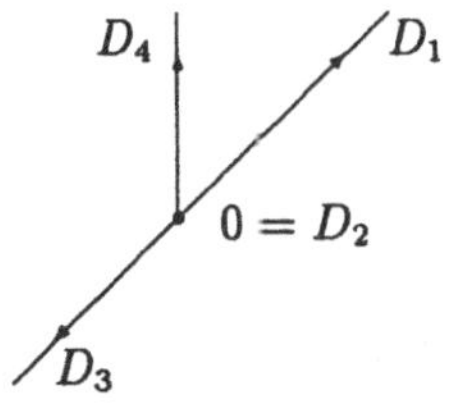

Let

$$D_{i_1}(\mathbf{v}) \prec D_{i_2}(\mathbf{v}) \prec \dots \prec D_{i_k}(\mathbf{v}) \tag{4.1.12}$$

be the list of all distinct nonzero complex numbers associated with $\mathbf{v} = (v_1, \dots, v_m)^t$ written in the order of increasing argument.

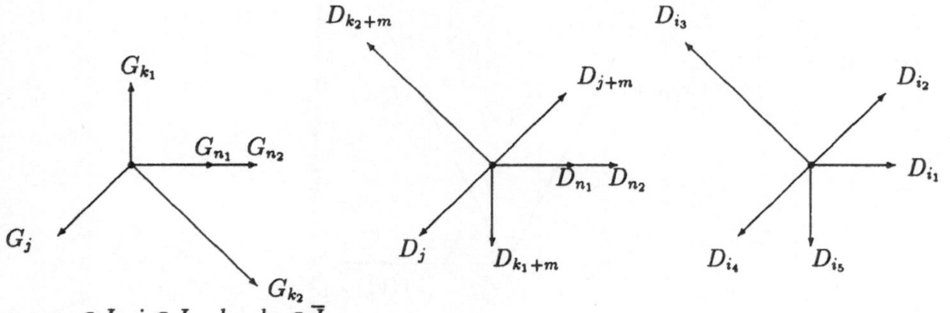

$n_1,\ n_2 \in \underline{I},\ j \in I_f,\ k_1,\ k_2 \in \overline{I}$

Condition (4.1.10) implies that

$$\arg \frac{D_{i_2}}{D_{i_1}} < \pi, \ldots, \arg \frac{D_{i_1}}{D_{i_k}} < \pi. \tag{4.1.13}$$

Let g be a restriction of f on a two dimensional cross-section of $\mathbf{B}^m$, i.e., $g : \mathbf{R}^2 \to \mathbf{C}$ where

$$g(\mathbf{y}) = g(y_1, y_2) = f(v_1, \ldots, v_{l_1-1}, y_1, v_{l_1+1}, \ldots, v_{l_2-1}, y_2, v_{l_2+1}, \ldots, v_m),$$

and

$$l_j = i_j \quad \text{if} \quad 1 \le i_j \le m, \quad \text{and} \quad l_j = i_j - m \quad \text{if} \quad m+1 \le i_j \le 2m, \quad j = 1, 2.$$

The Jacobian of g at $(y_1, y_2) = (v_{l_1}, v_{l_2})$ is

$$\begin{vmatrix} \operatorname{Re} \frac{\partial g}{\partial y_1}(\mathbf{y}) & \operatorname{Re} \frac{\partial g}{\partial y_2}(\mathbf{y}) \\ \operatorname{Im} \frac{\partial g}{\partial y_1}(\mathbf{y}) & \operatorname{Im} \frac{\partial g}{\partial y_2}(\mathbf{y}) \end{vmatrix} = \begin{vmatrix} \operatorname{Re} G_{l_1}(\mathbf{v}) & \operatorname{Re} G_{l_2}(\mathbf{v}) \\ \operatorname{Im} G_{l_1}(\mathbf{v}) & \operatorname{Im} G_{l_2}(\mathbf{v}) \end{vmatrix}.$$

Due to (4.1.12) the Jacobian does not vanish. For $j = 1,\ 2$ let

$$0 \le \epsilon_j \text{ if } 1 \le i_j \le m, \text{ and } \epsilon_j \le 0 \text{ if } m+1 \le i_j \le 2m.$$

There exists a positive scalar δ such that when $|\epsilon_j| \le \delta$, $j = 1,\ 2$ one has

1. $0 \le y_j + \epsilon_j \le 1$,

2. $$\begin{vmatrix} \operatorname{Re} \frac{\partial g}{\partial y_1}(y_1 + \epsilon_1, y_2 + \epsilon_2) & \operatorname{Re} \frac{\partial g}{\partial y_2}(y_1 + \epsilon_1, y_2 + \epsilon_2) \\ \operatorname{Im} \frac{\partial g}{\partial y_1}(y_1 + \epsilon_1, y_2 + \epsilon_2) & \operatorname{Im} \frac{\partial g}{\partial y_2}(y_1 + \epsilon_1, y_2 + \epsilon_2) \end{vmatrix} \ne 0.$$

We now invoke results valid for the two dimensional case. For each ϵ_1 and ϵ_2 that satisfy the above conditions the image of the two dimensional box with the vertices

$$\left\{(y_1, y_2)^t,\ (y_1 + \epsilon_1, y_2)^t, (y_1 + \epsilon_1, y_2 + \epsilon_2)^t, (y_1, y_2 + \epsilon_2)^t\right\}$$

is a four cornered convex polytope in the complex plane. The point $f(\mathbf{v})$ is a vertex of the polytope, and two of its edges are parallel to $D_{i_1}(\mathbf{v})$ and $D_{i_2}(\mathbf{v})$ respectively. We have just shown that the polytope is covered by $f(\mathbf{B}^m)$.

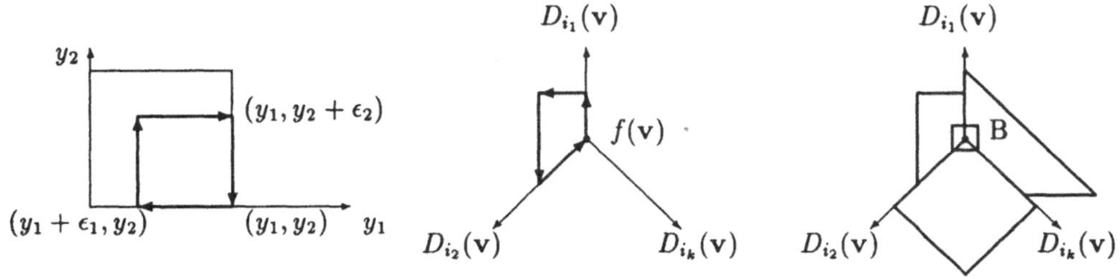

Repetition of these arguments for the pairs $\{D_{i_2}(\mathbf{v}), D_{i_3}(\mathbf{v})\}, \ldots, \{D_{i_k}(\mathbf{v}), D_{i_1}(\mathbf{v})\}$ shows the existence of a small two dimensional box B centered at $g(\mathbf{y}) = f(\mathbf{v})$ and covered by $f(\mathbf{B}^m)$.

Lemma 4.1.6 *Let* $\mathbf{v}$, $\mathbf{v}'$ *be neighboring vertices of* $\mathbf{B}^m$. *Suppose that* $\mathbf{v} \notin \left\{\mathbf{v}_p^1, \ldots, \mathbf{v}_p^{2m}\right\}$. *Let* $\mathbf{x}$ *be an interior point of the edge generated by* $\mathbf{v}$ *and* $\mathbf{v}'$, *i.e.,*

$$\mathbf{x} = \lambda \mathbf{v} + (1 - \lambda)\mathbf{v}', \quad 0 < \lambda < 1.$$

There exists a small square centered at $\mathbf{x}$ *and covered by* $f(\mathbf{B}^m)$.

Proof: To simplify the exposition we assume that

$$\mathbf{v} = (0, v_2, \ldots, v_m), \text{ and } \mathbf{v}' = (1, v_2, \ldots, v_m).$$

The condition $\mathbf{v} \notin \left\{\mathbf{v}_p^1, \ldots, \mathbf{v}_p^{2m}\right\}$ implies the existence of the indices i and j such that

$$(-1)^{v_i} G_i(\mathbf{v}) \prec f(\mathbf{v}') - f(\mathbf{v}) = G_1(\mathbf{v}) \prec (-1)^{v_j} G_j(\mathbf{v}).$$

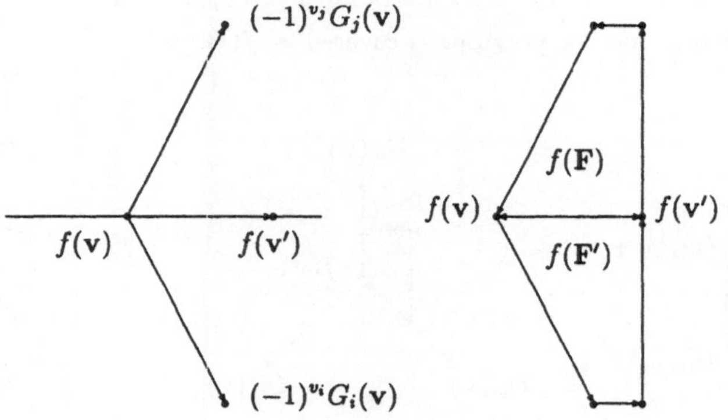

First consider the two dimensional face

$$\mathbf{F} = \{(x_1, v_2, \ldots, v_{j-1}, x_j, v_{j+1}, \ldots, v_m)\}.$$

The complex numbers $f(\mathbf{v})$ and $f(\mathbf{v}')$ are consecutive vertices of the convex polygon $f(\mathbf{F})$. Next consider the two dimensional face

$$\mathbf{F}' = \{(x_1, v_2, \ldots, v_{i-1}, x_i, v_{i+1}, \ldots, v_m)\}.$$

The complex numbers $f(\mathbf{v}')$ and $f(\mathbf{v})$ are consecutive vertices of the convex polygon $f(\mathbf{F}')$. The proof of the lemma is now completed.

Finally we show that an ϵ "neighborhood" of ∂conv $f(\mathbf{B}^m)$ is covered by $f(\mathbf{B}^m)$.

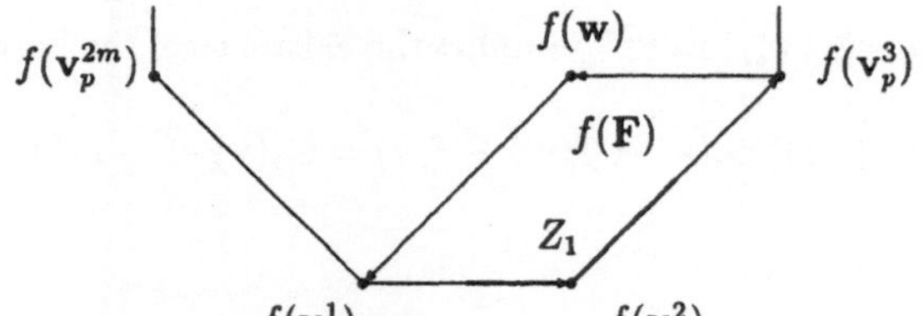

Lemma 4.1.7 *There exists a positive scalar ϵ such that each $z \in$ conv $f(\mathbf{B}^m)$ located in the ϵ vicinity of the boundary ∂conv $f(\mathbf{B}^m)$ is covered by $f(\mathbf{B}^m)$.*

Proof: We first prove the statement for a special subset of the boundary Z_1 defined by

$$\left\{f(\mathbf{v}_p^2)+\alpha\left[f(\mathbf{v}_p^3)-f(\mathbf{v}_p^2)\right],\ 0\le\alpha\le\frac{1}{2}\right\}\cup\left\{f(\mathbf{v}_p^2)+\alpha\left[f(\mathbf{v}_p^1)-f(\mathbf{v}_p^2)\right],\ 0\le\alpha\le\frac{1}{2}\right\}.$$

Consider the four vertices $\mathbf{v}_p^1$, $\mathbf{v}_p^2$, $\mathbf{v}_p^3$, and $\mathbf{w}=(0,1,0,\ldots,0)^t$ of the two dimensional face

$$\mathbf{F}=\left\{(x_1,x_2,0,\ldots,0)^t\right\}.$$

The complex numbers $f(\mathbf{v}_p^1)$, $f(\mathbf{v}_p^2)$, $f(\mathbf{v}_p^3)$, and $f(\mathbf{w})$ are consecutive vertices of the convex polygon $f(\mathbf{F})$. This shows the existence of a positive ϵ_1 so that every $z\in\text{conv } f(\mathbf{B}^m)$ in ϵ_1 vicinity of Z_1 is covered by $f(\mathbf{B}^m)$. Analogously define subsets of the boundary Z_i and obtain positive scalars ϵ_i, $i=2,\ldots,2m$. Observe that $\partial\text{conv } f(\mathbf{B}^m)=\cup_{i=1}^{2m}Z_i$, and define $\epsilon=\min\{\epsilon_1,\ldots,\epsilon_{2m}\}$.

Now we are ready to prove (4.1.9). Suppose the opposite, i.e., there exists $z_0\in$ conv $f(\mathbf{B}^m)$ such that $d(z_0,f(\mathbf{B}^m))>0$. Let $\mathcal{F}_2$ be the union of all the two dimensional faces of the box $\mathbf{B}^m$, i.e.,

$$\mathcal{F}_2=\left\{\bigcup\mathbf{F}\ :\ \mathbf{F}\text{ is a two dimensional face of }\mathbf{B}^m\right\}.$$

Clearly $\mathcal{F}_2\subset\mathbf{B}^m$, and $d(z_0,f(\mathcal{F}_2))\ge d(z_0,f(\mathbf{B}^m))>0$. Let $\mathbf{x}\in\mathcal{F}_2$ such that $|z_0-f(\mathbf{x})|=d(z_0,f(\mathcal{F}_2))>0$. Then $z=f(\mathbf{x})$ belongs to the image of a one dimensional edge of the box, i.e.,

$$z=\lambda f(\mathbf{v})+(1-\lambda)f(\mathbf{v}'),\quad 0\le\lambda\le1.$$

According to Lemma 4.1.7 $\mathbf{v}$, and $\mathbf{v}'$ may not both be principal vertices. If $\mathbf{v}\notin\left\{v_p^1,\ldots,v_p^{2m}\right\}$, then, due to Lemma 4.1.6, either $\lambda=0$, or $\lambda=1$. According to Lemma 4.1.5 one has $\lambda\neq1$, and $\mathbf{v}'$ must be a principal vertex. This contradicts Lemma 4.1.7, and completes the proof.

4.2. Image of a Box Under Multiaffine Transformation

In this section we do not impose orientation preserving conditions on a multiaffine mappings f, and provide a simple analytic description of the boundary of the image of the box $\mathbf{B}^m$ under f. If $\mathbf{x}$ is an interior point of $\mathbf{B}^m$, i.e., $i\in I_f(\mathbf{x})$, $i=1,\ldots,m$, and $f(\mathbf{x})\in\partial f(\mathbf{B}^m)$, then all the partial derivatives $G_i(\mathbf{x})$ of f at $\mathbf{x}$ must be proportional. Generally, if $f(\mathbf{x})\in\partial f(\mathbf{B}^m)$, then the complex numbers

$$\left\{G_j(\mathbf{x}),\ j\in\underline{I}(\mathbf{x})\ \right\},\ \left\{G_i(\mathbf{x}),\ i\in I_f(\mathbf{x})\ \right\},\ \left\{-G_k(\mathbf{x}),\ k\in\overline{I}(\mathbf{x})\right\}$$

should belong to a half plane of the complex plane **C**. Keeping in mind that a half plane is defined by a non zero complex number below we define *a principal point* associated with the unit box $\mathbf{B}^m$ and a mapping $f : \mathbf{B}^m \to \mathbf{C}$. A principal point associated with a box **B** and a mapping $f : \mathbf{B} \to \mathbf{C}$ can be defined analogously.

Definition 4.2.1 $\mathbf{x} \in \mathbf{B}^m$ *is a principal point if there exists a non zero complex number* G *such that:*

$$G_i(\mathbf{x}) \asymp G \text{ for each } i \in I_f(\mathbf{x}),\ G_i(\mathbf{x}) \neq 0 \qquad (4.2.1)$$

$$G_i(\mathbf{x}) \preceq G \preceq G_k(\mathbf{x}) \text{ for each } G_i(\mathbf{x}) \neq 0,\ i \in \underline{I}(\mathbf{x}),\ G_k(\mathbf{x}) \neq 0,\ k \in \overline{I}(\mathbf{x}). \qquad (4.2.2)$$

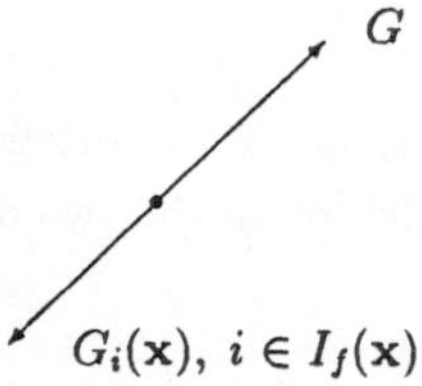

$G_i(\mathbf{x}),\ i \in I_f(\mathbf{x})$

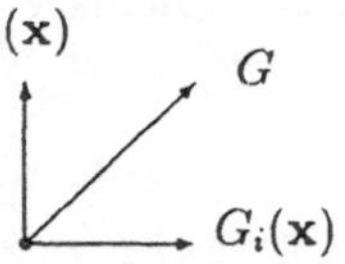

$G_i(\mathbf{x}) \neq 0,\ i \in \underline{I}(\mathbf{x});\ G_k(\mathbf{x}) \neq 0,\ k \in \overline{I}(\mathbf{x})$

The main result of the section is given next.

Theorem 4.2.1 If $z \in \partial f(\mathbf{B}^m)$, then there exists a principal point $\mathbf{x}$ such that $z = f(\mathbf{x})$.

The proof of this theorem is given in the next subsection. The theorem shows that images of principal points cover the boundary of $f(\mathbf{B}^m)$. The set of principal points is *not* the smallest subset of the box whose image covers the boundary of the image of the box. However this set admits a simple analytic description which is given below.

1. Choose the sets of indices $\underline{I}$, I_f, and $\overline{I}$. Set $x_i = 0$ for $i \in \underline{I}$, and $x_i = 1$ for $i \in \overline{I}$.

2. If $|I_f| \geq 2$ (where $|I_f|$ stands for the number of elements of the set I_f), then solve the system of nonlinear equations with respect to $\mathbf{x}$

$$G_i(\mathbf{x}) \asymp G_k(\mathbf{x}) \text{ for each } i, k \in I_f \text{ such that } G_i(\mathbf{x}) \neq 0,\ G_k(\mathbf{x}) \neq 0. \qquad (4.2.3)$$

3. Generically (4.2.3) defines a manifold of dimension one. For each point $\mathbf{x}$ of the manifold define G as a nonzero complex number satisfying (4.2.1).

4. Check that (4.2.2) holds.

Characterization of principal points requires to solve a system of nonlinear equations. In many important special cases the description of principal points is available. Principal points associated with special examples of multiaffine polynomial families are described in Section 4.4 and Section 4.5.

4.2.1. Proof of Main Result

Lemma 4.2.1 *If $f(\mathbf{x}) \in \partial f(\mathbf{B}^m)$, then there exists a nonzero complex number G such that*

$$D_i(\mathbf{x}) \preceq G \text{ for each } D_i(\mathbf{x}) \neq 0. \tag{4.2.4}$$

Proof: Let

$$D_{i_1} \prec D_{i_2} \prec \ldots \prec D_{i_k}$$

be the list of all distinct nonzero complex numbers associated with $\mathbf{x}$ written in the order of increasing argument (see (4.1.11), page 69). Suppose that the conclusion of the lemma does not hold, i.e.,

$$\arg \frac{D_{i_2}}{D_{i_1}} < \pi, \ \ldots, \arg \frac{D_{i_1}}{D_{i_k}} < \pi. \tag{4.2.5}$$

In what follows we show that (4.2.5) leads to a contradiction. Let g be a restriction of f on a two dimensional cross-section of $\mathbf{B}^m$, i.e., $g \ : \ \mathbf{R}^2 \to \mathbf{C}$ where

$$g(\mathbf{y}) = g(y_1, y_2) = f(x_1, \ldots, x_{l_1-1}, y_1, x_{l_1+1}, \ldots, x_{l_2-1}, y_2, x_{l_2+1}, \ldots, x_m),$$

and

$$l_j = i_j \quad \text{if} \quad 1 \le i_j \le m, \quad \text{and} \quad l_j = i_j - m \quad \text{if} \quad m+1 \le i_j \le 2m, \ \ j = 1, 2.$$

Repetition of the arguments given in the proof of Lemma 4.1.5 leads to a contradiction and completes the proof.

Relations (4.1.11) and Definition 4.2.1 show that $\mathbf{x} \in \mathbf{B}^m$ is a principal point if and only if there exists a nonzero complex number G such that (4.2.4) holds. The proof of Theorem 4.2.1 follows now from Lemma 4.2.1.

4.3. Robust Stability of Multiaffine Polynomial Families

The remainder of the chapter is concerned with Hurwitz stability of the polynomial family

$$\mathbf{P} = \{\mathbf{p}(s, \mathbf{x}) \ : \ \mathbf{p}(s, \mathbf{x}) = a_0(\mathbf{x}) + a_1(\mathbf{x})s + \ldots + a_{n-1}(\mathbf{x})s^{n-1} + a_n(\mathbf{x})s^n, \ \mathbf{x} \in \mathbf{B}\},$$

where $a_i(\mathbf{x})$, $i = 0, 1, \ldots, n$ are multiaffine mappings from $\mathbf{R}^m$ to $\mathbf{R}$.

Definition 4.3.1 *For a given frequency* w *the set of principal points of the polynomial family* $\mathbf{P}$ *at* w *is defined as follows*

$$\mathbf{X_p}(w) = \left\{ \ \mathbf{x} \in \mathbf{B} \ : \ \mathbf{x} \textit{ is a principal point of } f(\mathbf{x}) = \mathbf{p}(jw, \mathbf{x}) \ \right\}. \qquad (4.3.1)$$

The set of principal points of the family $\mathbf{P}$ *is defined by* $\mathbf{X_p} = \bigcup_{0 \leq w} \mathbf{X_p}(w)$.

The "zero exclusion" criterion along with Theorem 4.2.1 yield the following stability criterion.

Theorem 4.3.1 Assume that for some $\mathbf{x}^0 \in \mathbf{B}$ the polynomial $\mathbf{p}(s, \mathbf{x}^0)$ is stable, and $a_0(\mathbf{x}) \neq 0$, $a_n(\mathbf{x}) \neq 0$ for each $\mathbf{x} \in \mathbf{B}$. The polynomial family $\mathbf{P}$ is stable if and only if the polynomials $\{\mathbf{p}(s, \mathbf{x}) \ : \ \mathbf{x} \in \mathbf{X_p}\}$ are stable.

The frequency domain version of the theorem is given next.

Theorem 4.3.2 Assume that for some $\mathbf{x}^0 \in \mathbf{B}$ the polynomial $\mathbf{p}(s, \mathbf{x}^0)$ is stable, and $a_0(\mathbf{x}) \neq 0$, $a_n(\mathbf{x}) \neq 0$ for each $\mathbf{x} \in \mathbf{B}$. The polynomial family $\mathbf{P}$ is stable if and only if

$$0 \notin \{\mathbf{p}(jw, \mathbf{x}) \ : \ \mathbf{x} \in \mathbf{X_p}(w)\} \ \text{ for each } w > 0. \qquad (4.3.2)$$

Although generically the set $\mathbf{X_p}(w)$ depends on w, in the next section we present a problem where $\mathbf{X_p}(w) = \mathbf{X_p}$ for each $0 \leq w$.

4.4. Cascade of Uncertain Blocks

This section is concerned with a closed loop with a cascade of first order blocks involving uncertainty in time constants:

$$x_i \in [\underline{x}_i, \overline{x}_i], \ i = 1, \ldots, m.$$

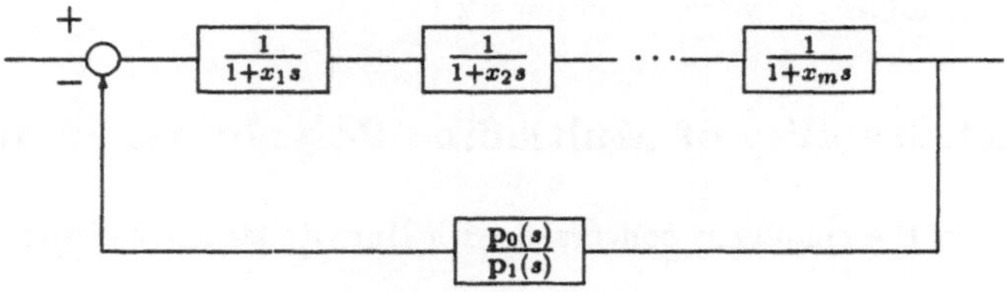

The characteristic polynomials of the closed loop are

$$\mathbf{P} = \left\{ \mathbf{p}(s, \mathbf{x}) = \mathbf{p}_0(s) + \mathbf{p}_1(s) \prod_{k=1}^{m} (1 + x_k s), \ \mathbf{x} \in \mathbf{B} \right\}. \qquad (4.4.1)$$

To avoid the "degree dropping" and trivial situations we assume that $0 \notin [\underline{x}_i, \overline{x}_i]$, and $\mathbf{p}_1(\boldsymbol{j}w) \neq 0$ for each real w.

Let $w > 0$ be fixed. Define $f : \mathbf{B} \to \mathbf{C}$ by $f(\mathbf{x}) = \mathbf{p}(\boldsymbol{j}w, \mathbf{x})$. Then

$$G_i(\mathbf{x}) = \boldsymbol{j}w\mathbf{p}_1(\boldsymbol{j}w) \prod_{k \neq i} (1 + x_k \boldsymbol{j}w), \; i = 1, \ldots, m. \tag{4.4.2}$$

Definition 4.4.1 *Principal Edges.*
A principal edge is an edge of $\mathbf{B}$

$$\{\mathbf{x} : \mathbf{x} \in \mathbf{B}, \; \underline{x}_k \leq x_k \leq \overline{x}_k, \; i \notin I_f \text{ when } i \neq k\}$$

such that the coordinates x_i with corresponding intervals $[\underline{x}_i, \overline{x}_i]$ not intersecting with $[\underline{x}_k, \overline{x}_k]$ are determined either by

$$|x_i - x_k| = \max\{|x - x_k| \; : \; x \in [\underline{x}_i, \overline{x}_i]\} \text{ for each } i \neq k,$$

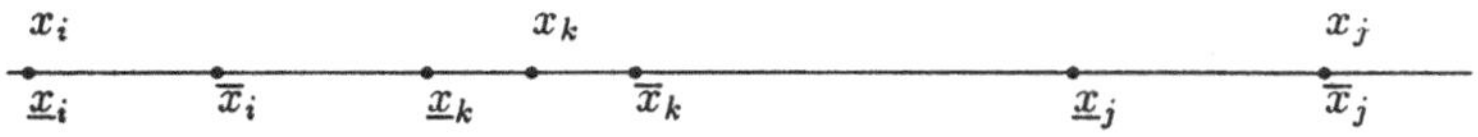

or by

$$|x_i - x_k| = \min\{|x - x_k| \; : \; x \in [\underline{x}_i, \overline{x}_i]\} \text{ for each } i \neq k.$$

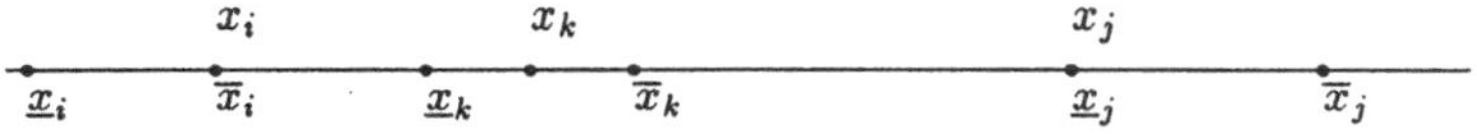

Definition 4.4.2 *Principal Segments.*
Let σ be a common point of more than one open interval $(\underline{x}_i, \overline{x}_i)$. That is there exists a set of indices $I_\sigma = \{i_1, \ldots, i_k\}$, $k \geq 2$ such that

$$\sigma \in (\underline{x}_i, \overline{x}_i) \text{ if } i \in I_\sigma, \text{ and } \sigma \notin (\underline{x}_i, \overline{x}_i) \text{ if } i \notin I_\sigma.$$

A principal segment of $\mathbf{B}$ *is defined by*

$$\left\{ \mathbf{x} : \mathbf{x} \in \mathbf{B}, \; \begin{array}{lll} x_i = \sigma & \text{if} & i \in I_\sigma, \\ x_i = \underline{x}_i & \text{if} & i \notin I_\sigma, \text{ and } \sigma \leq \underline{x}_i; \\ x_i = \overline{x}_i & \text{if} & i \notin I_\sigma, \text{ and } \sigma \geq \overline{x}_i. \end{array} \right\}$$

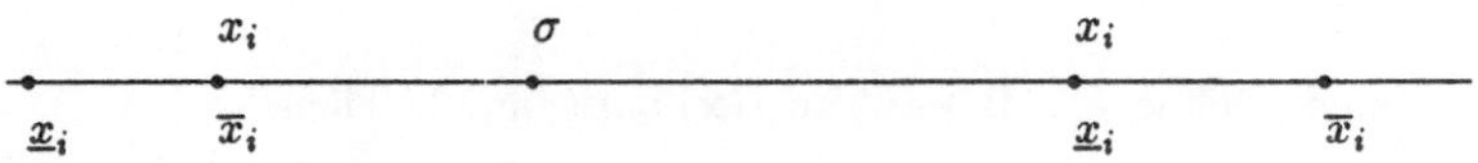

The next statement describes the set of principal points $\mathbf{X_p}$.

Theorem 4.4.1 Principal edges and principal segments form the set of principal points $\mathbf{X_p}$.

The proof of the theorem is given in the following subsection.

Corollary 4.4.1 *The polynomial family (4.4.1) is stable if and only if all principal edges and principal segments are stable.*

The analogous (but computationally more demanding) result can be found in [FDB], Theorem 3.1.

Next we provide a number of specific examples in which explicit description of principal points is available.

Example 4.4.1 (See [SEPB] , [SEP1], [Po]). Non overlapping intervals.
If $\underline{x}_1 < \overline{x}_1 < \underline{x}_2 < \overline{x}_2 \ldots < \underline{x}_m < \overline{x}_m$, then the polynomial family (4.4.1) is stable if and only if $2m$ principal edges of the box $\mathbf{B}$

$$\{\mathbf{x} \; : \; \underline{x}_k \leq x_k \leq \overline{x}_k, \; x_i = \underline{x}_i, \; 1 \leq i < k; \; x_i = \overline{x}_i, \; k < i \leq m\}$$

$$\{\mathbf{x} \; : \; \underline{x}_k \leq x_k \leq \overline{x}_k, \; x_i = \overline{x}_i, \; 1 \leq i < k; \; x_i = \underline{x}_i, \; k < i \leq m\}$$

are stable.

Example 4.4.2 Identical intervals.
If $[\underline{x}_i, \overline{x}_i] = [\underline{x}, \overline{x}]$, then the polynomial family (4.4.1) is stable if and only if m principal edges of the box

$$\{\mathbf{x} \; : \; \underline{x} \leq x_1 \leq \overline{x}, \; x_2 = \underline{x}, \; x_3 = \underline{x}, \; \ldots, x_m = \underline{x}\}$$
$$\{\mathbf{x} \; : \; \underline{x} \leq x_1 \leq \overline{x}, \; x_2 = \overline{x}, \; x_3 = \underline{x}, \; \ldots, x_m = \underline{x}\}$$
$$\{\mathbf{x} \; : \; \underline{x} \leq x_1 \leq \overline{x}, \; x_2 = \overline{x}, \; x_3 = \overline{x}, \; \ldots, x_m = \underline{x}\}$$
$$\ldots$$
$$\{\mathbf{x} \; : \; \underline{x} \leq x_1 \leq \overline{x}, \; x_2 = \overline{x}, \; x_3 = \overline{x}, \; \ldots, x_m = \overline{x}\}$$

and one principal segment

$$\{\mathbf{x} \ : \ x_1 = x_2 = \ldots = x_m = \sigma, \ \underline{x} \le \sigma \le \overline{x}\}$$

are stable.

While the principal point technique provides simple necessary and sufficient stability conditions, even in a particular case of this simple example the Mapping Theorem is excessively conservative. Indeed, consider the polynomial family

$$\mathbf{P} = \left\{ \mathbf{p}(s, \mathbf{x}) = \prod_{k=1}^{m} (1 + x_k s) \, , \ 0 < \underline{x} \le x_i \le \overline{x} \right\}. \tag{4.4.3}$$

It is easy to see that the family (4.4.3) is stable. On the other hand the vertices of the box are mapped under $\mathbf{p}(\boldsymbol{j}w, \mathbf{x})$ into the $m + 1$ points

$$(1 + \boldsymbol{j}w\underline{x})^m \left(\frac{1 + \boldsymbol{j}w\overline{x}}{1 + \boldsymbol{j}w\underline{x}} \right)^i, \ i = 0, \ldots, m.$$

If for some $w > 0$ one has $\arg \dfrac{1 + \boldsymbol{j}w\overline{x}}{1 + \boldsymbol{j}w\underline{x}} > \dfrac{\pi}{m}$, then the images of the vertices encircle the origin, the convexification of the value set provided by the Mapping Theorem contains the origin, and the Mapping Theorem is inconclusive.

Example 4.4.3 Nested intervals.
If $[\underline{x}_1, \overline{x}_1] \subseteq [\underline{x}_2, \overline{x}_2] \subseteq \ldots \subseteq [\underline{x}_m, \overline{x}_m]$, then the polynomial family (4.4.1) is stable if and only if all the exposed edges of the box, and $2m - 3$ principal segments

$$\{\mathbf{x} \ : \ x_1 = x_2 = \ldots = x_m = \sigma, \ \underline{x}_1 \le \sigma \le \overline{x}_1\}$$

$$\{\mathbf{x} \ : \ x_1 = \underline{x}_1, \ x_2 = \underline{x}_2, \ \ldots, x_k = \underline{x}_k, \ x_{k+1} = \ldots = x_m = \sigma, \ \underline{x}_{k+1} \le \sigma \le \underline{x}_k\}$$

$$\ldots$$

$$\{\mathbf{x} \ : \ x_1 = \overline{x}_1, \ x_2 = \overline{x}_2, \ \ldots, x_k = \overline{x}_k, \ x_{k+1} = \ldots = x_m = \sigma, \ \overline{x}_k \le \sigma \le \overline{x}_{k+1}\}$$

$(k = 1, \ldots, m - 2)$ are stable.

Example 4.4.4 Interlacing intervals.

$$\underline{x}_1 < \underline{x}_2 < \overline{x}_1 < \underline{x}_3 < \overline{x}_2 < \ldots < \underline{x}_m < \overline{x}_{m-1} < \overline{x}_m.$$

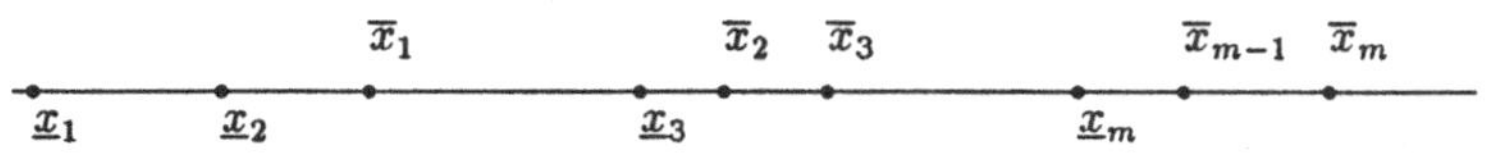

The polynomial family (4.4.1) is stable if and only if $8m-8$ principal edges

$$\{\mathbf{x} \,:\, \underline{x}_1 \le x_1 \le \overline{x}_1,\ x_2 = \underline{x}_2,\ x_3 = \underline{x}_3, x_4 = \underline{x}_4, \ldots, x_m = \underline{x}_m\}$$
$$\{\mathbf{x} \,:\, \underline{x}_1 \le x_1 \le \overline{x}_1,\ x_2 = \overline{x}_2,\ x_3 = \underline{x}_3, x_4 = \underline{x}_4, \ldots, x_m = \underline{x}_m\}$$
$$\{\mathbf{x} \,:\, \underline{x}_1 \le x_1 \le \overline{x}_1,\ x_2 = \underline{x}_2,\ x_3 = \overline{x}_3, x_4 = \overline{x}_4, \ldots, x_m = \overline{x}_m\}$$
$$\{\mathbf{x} \,:\, \underline{x}_1 \le x_1 \le \overline{x}_1,\ x_2 = \overline{x}_2,\ x_3 = \overline{x}_3, x_4 = \overline{x}_4, \ldots, x_m = \overline{x}_m\}$$
$$\{\mathbf{x} \,:\, \underline{x}_m \le x_m \le \overline{x}_m,\ x_{m-1} = \underline{x}_{m-1},\ x_1 = \underline{x}_1, x_2 = \underline{x}_2, \ldots, x_{m-2} = \underline{x}_{m-2}\}$$
$$\{\mathbf{x} \,:\, \underline{x}_m \le x_m \le \overline{x}_m,\ x_{m-1} = \overline{x}_{m-1},\ x_1 = \underline{x}_1, x_2 = \underline{x}_2, \ldots, x_{m-2} = \underline{x}_{m-2}\}$$
$$\{\mathbf{x} \,:\, \underline{x}_m \le x_m \le \overline{x}_m,\ x_{m-1} = \underline{x}_{m-1},\ x_1 = \overline{x}_1, x_2 = \overline{x}_2, \ldots, x_{m-2} = \overline{x}_{m-2}\}$$
$$\{\mathbf{x} \,:\, \underline{x}_m \le x_m \le \overline{x}_m,\ x_{m-1} = \overline{x}_{m-1},\ x_1 = \overline{x}_1, x_2 = \overline{x}_2, \ldots, x_{m-2} = \overline{x}_{m-2}\}$$

$$\left\{\mathbf{x} \,:\, \underline{x}_k \le x_k \le \overline{x}_k, \begin{array}{cc} x_{k-1} = \underline{x}_{k-1} & x_{k+1} = \underline{x}_{k+1} \\ x_i = \underline{x}_i,\ 1 \le i < k-1 & x_j = \overline{x}_j,\ k+1 < j \le m \end{array}\right\}$$
$$\left\{\mathbf{x} \,:\, \underline{x}_k \le x_k \le \overline{x}_k, \begin{array}{cc} x_{k-1} = \underline{x}_{k-1} & x_{k+1} = \overline{x}_{k+1} \\ x_i = \underline{x}_i,\ 1 \le i < k-1 & x_j = \overline{x}_j,\ k+1 < j \le m \end{array}\right\}$$
$$\left\{\mathbf{x} \,:\, \underline{x}_k \le x_k \le \overline{x}_k, \begin{array}{cc} x_{k-1} = \overline{x}_{k-1} & x_{k+1} = \underline{x}_{k+1} \\ x_i = \underline{x}_i,\ 1 \le i < k-1 & x_j = \overline{x}_j,\ k+1 < j \le m \end{array}\right\}$$
$$\left\{\mathbf{x} \,:\, \underline{x}_k \le x_k \le \overline{x}_k, \begin{array}{cc} x_{k-1} = \overline{x}_{k-1} & x_{k+1} = \overline{x}_{k+1} \\ x_i = \underline{x}_i,\ 1 \le i < k-1 & x_j = \overline{x}_j,\ k+1 < j \le m \end{array}\right\}$$

$$k = 2, \ldots, m-1$$

$$\left\{\mathbf{x} \,:\, \underline{x}_k \le x_k \le \overline{x}_k, \begin{array}{cc} x_{k-1} = \underline{x}_{k-1} & x_{k+1} = \underline{x}_{k+1} \\ x_i = \overline{x}_i,\ 1 \le i < k-1 & x_j = \underline{x}_j,\ k+1 < j \le m \end{array}\right\}$$
$$\left\{\mathbf{x} \,:\, \underline{x}_k \le x_k \le \overline{x}_k, \begin{array}{cc} x_{k-1} = \underline{x}_{k-1} & x_{k+1} = \overline{x}_{k+1} \\ x_i = \overline{x}_i,\ 1 \le i < k-1 & x_j = \underline{x}_j,\ k+1 < j \le m \end{array}\right\}$$
$$\left\{\mathbf{x} \,:\, \underline{x}_k \le x_k \le \overline{x}_k, \begin{array}{cc} x_{k-1} = \overline{x}_{k-1} & x_{k+1} = \underline{x}_{k+1} \\ x_i = \overline{x}_i,\ 1 \le i < k-1 & x_j = \underline{x}_j,\ k+1 < j \le m \end{array}\right\}$$
$$\left\{\mathbf{x} \,:\, \underline{x}_k \le x_k \le \overline{x}_k, \begin{array}{cc} x_{k-1} = \overline{x}_{k-1} & x_{k+1} = \overline{x}_{k+1} \\ x_i = \overline{x}_i,\ 1 \le i < k-1 & x_j = \underline{x}_j,\ k+1 < j \le m \end{array}\right\}$$

and $m-1$ principal segments

$$\{\mathbf{x} \,:\, x_1 = x_2 = \sigma,\ \underline{x}_2 \le \sigma \le \overline{x}_1,\ x_3 = \underline{x}_3,\ x_4 = \underline{x}_4, \ldots, x_m = \underline{x}_m\}$$
$$\{\mathbf{x} \,:\, x_2 = x_3 = \sigma,\ \underline{x}_3 \le \sigma \le \overline{x}_2,\quad x_1 = \overline{x}_1,\ x_4 = \underline{x}_4, \ldots, x_m = \underline{x}_m\}$$
$$\ldots$$
$$\{\mathbf{x} \,:\, x_{m-1} = x_m = \sigma,\ \underline{x}_m \le \sigma \le \overline{x}_{m-1},\ x_1 = \overline{x}_1,\ x_2 = \overline{x}_2, \ldots, x_{m-2} = \overline{x}_{m-2}\}$$

are stable.

Remark 4.4.1 An elegant result concerning evaluation of the robust gain margin for a cascade of uncertain blocks was recently reported by Kiselev and Polyak [KP]. The work deals with the family of characteristic polynomials

$$\left\{ \mathbf{p}(s,\mathbf{x}) = k\mathbf{p}_0(s) + \mathbf{p}_1(s) \prod_{i=1}^{m} (1 + x_i s), \ \mathbf{x} \in \mathbf{B}, \quad \begin{matrix} 0 \leq k \leq \overline{k} \\ x_i \in [\underline{x}_i, \overline{x}_i] \\ i = 1, \dots, m \end{matrix} \right\}. \tag{4.4.4}$$

The robust gain margin k^* is defined as

$$k^* = \sup \left\{ \overline{k} \text{ all polynomials (4.4.4) are stable} \right\}.$$

A simple convexity argument reduces evaluation of k^* to minimization of a real valued function $k(x)$ over a finite interval $[\underline{x}, \overline{x}]$, where $\underline{x} = \min_i \underline{x}_i$, and $\overline{x} = \max_i \overline{x}_i$.

4.4.1. Principal Edges and Principal Segments

We now proof Theorem 4.4.1. Let $\mathbf{x}$ be a principal point of the multiaffine mapping $f(\mathbf{x}) = \mathbf{p}(\boldsymbol{j}w, \mathbf{x})$ given by (4.4.1), and $w > 0$. If $k,\ i \in I_f(\mathbf{x})$, then

$$G_k(\mathbf{x}) \asymp G_i(\mathbf{x}), \text{ and } 0 = \operatorname{Im} \frac{G_i(\mathbf{x})}{G_k(\mathbf{x})} = \operatorname{Im} \frac{1 + \boldsymbol{j}wx_k}{1 + \boldsymbol{j}wx_i}. \tag{4.4.5}$$

The equation yields $x_k = x_i$. We next show that coordinates x_i, x_j with $i \in \underline{I}$, and $j \in \overline{I}$ obey the rules of Definition 4.4.1 and Definition 4.4.2.

Lemma 4.4.1 *Principal Edges*

Suppose that $\mathbf{x}$ *is a principal point and* $|I_f(\mathbf{x})| = 1$. *Then* $\mathbf{x}$ *belongs to a principal edge.*

Proof: Suppose that $\underline{x}_k < x_k < \overline{x}_k$, $\overline{x}_i < x_k$ and $x_k < \underline{x}_j$. If $x_i = \overline{x}_i$, and $x_j = \overline{x}_j$,

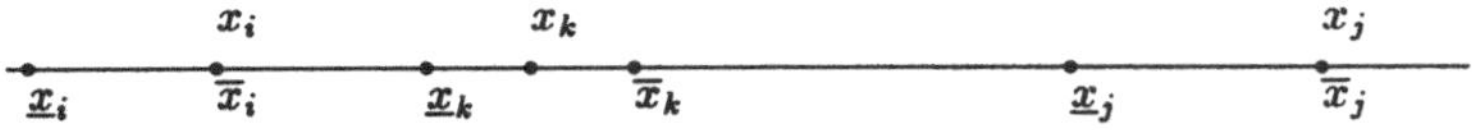

then $G_i \neq 0$, $G_j \neq 0$, and $G_j \prec G_k \prec G_i$. The only possible choices of G are αG_k, or $-\alpha G_k$, $\alpha > 0$. In both cases relations (4.2.2) are violated. Similar arguments show that (4.2.2) are violated if $x_i = \underline{x}_i$, and $x_j = \underline{x}_j$.

When f is a multiaffine function, $G_k(\mathbf{x})$ is a multiaffine function as well, and

$$G_k(\mathbf{x}^i(\gamma)) = G_k(x_1, \dots, x_i + \gamma, \dots, x_m)$$

is a linear function of γ. In particular $\arg G_k(\mathbf{x}^i(\gamma))$ is either a constant, or a strictly monotone function of γ.

Lemma 4.4.2 *Let $k,\ i \in I_f(\mathbf{x})$. If arg $G_k(\mathbf{x}^i(\gamma))$ is a strictly increasing function of γ, then $f(\mathbf{B}^m)$ covers a small square attached to $f(\mathbf{x})$ from the "right" of the line generated by $G_k(\mathbf{x})$.*

Proof: Let ϵ be a small positive number so that the two dimensional box centered at $\mathbf{x}$ with sides of size 2ϵ parallel to the axes x_i, x_k does not leave the box $\mathbf{B}^m$. When γ moves from $-\epsilon$ to ϵ the one dimensional intervals

$$\left[\mathbf{x}^i(\gamma) - \epsilon\mathbf{e}_k, \mathbf{x}^i(\gamma) + \epsilon\mathbf{e}_k\right]$$

sweep a small square attached to $f(\mathbf{x})$ from the "right".

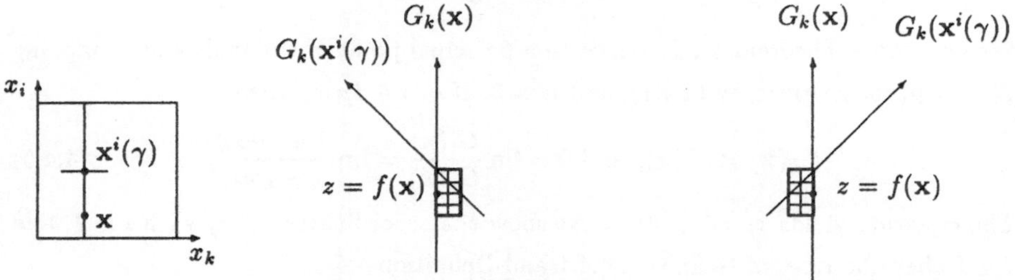

increasing arg $G_k(\mathbf{x}^i(\gamma))$ decreasing arg $G_k(\mathbf{x}^i(\gamma))$

This completes the proof.

A similar statement holds when arg $G_k(\mathbf{x}^i(\gamma))$ is a strictly decreasing function of γ. The next statement follows from Lemma 4.4.2.

Lemma 4.4.3 *Let $\mathbf{x}$ be a principal point with $i,\ k \in I_f(\mathbf{x})$.*

1. *If arg $G_k(\mathbf{x}^i(\gamma))$ is a strictly increasing function of γ, then $D_j(\mathbf{x}) \preceq G_k(\mathbf{x})$ for each nonzero $D_j(\mathbf{x})$.*

2. *If arg $G_k(\mathbf{x}^i(\gamma))$ is a strictly decreasing function of γ, then $D_j(\mathbf{x}) \succeq G_k(\mathbf{x})$ for each nonzero $D_j(\mathbf{x})$.*

Lemma 4.4.4 Principal Segments.
Suppose that $\mathbf{x}$ is a principal point and $|I_f(\mathbf{x})| \geq 2$. Then $\mathbf{x}$ belongs to a principal segment.

Proof: A straightforward verification shows that when $G_k(\mathbf{x})$ is given by (4.4.2) arg $G_k(\mathbf{x}^i(\gamma))$ is a strictly increasing function of γ. Application of Lemma 4.4.3 shows that there exists no $D_j(\mathbf{x})$ with $D_j(\mathbf{x}) \succ G_k(\mathbf{x})$.

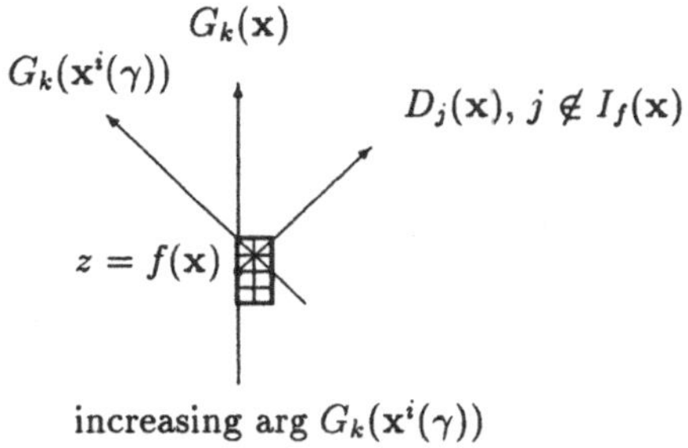

This relation along with relations (4.1.11) yields the following for $j \notin I_\sigma$

$$x_j = \underline{x}_j \text{ if } \sigma \leq \underline{x}_j, \text{ and } x_j = \overline{x}_j \text{ if } \sigma \geq \overline{x}_j.$$

This completes the proof.

Lemma 4.4.1 and Lemma 4.4.4 describe principal points $\mathbf{x}$ with $|I_f(\mathbf{x})| \geq 1$. We next show that if a principal point $\mathbf{x}$ is a vertex of the box (i.e., $|I_f(\mathbf{x})| = 0$), then $\mathbf{x}$ always belongs either to a principal edge, or to a principal segment. This will complete the proof of Theorem 4.4.1.

We denote the closed sets of principal edges and principal segments by $\mathcal{E}_\mathbf{p}$ and $\mathcal{S}_\mathbf{p}$ respectively, and observe that

$$\partial f(\mathbf{B}) - f(\mathcal{E}_\mathbf{p} \cup \mathcal{S}_\mathbf{p}) \subseteq \left\{f\left(\mathbf{v}^1\right), \ldots, f\left(\mathbf{v}^{2^m}\right)\right\}$$

where $\mathbf{v}^i$ is a vertex of $\mathbf{B}$, and $\partial f(\mathbf{B}) - f(\mathcal{E}_\mathbf{p} \cup \mathcal{S}_\mathbf{p})$ denotes the set of elements of $\partial f(\mathbf{B})$ that do not belong to $f(\mathcal{E}_\mathbf{p} \cup \mathcal{S}_\mathbf{p})$. In what follows we show that $\partial f(\mathbf{B}) - f(\mathcal{E}_\mathbf{p} \cup \mathcal{S}_\mathbf{p}) = \emptyset$. Suppose the opposite, i.e., let $z_0 \in \partial f(\mathbf{B}) - f(\mathcal{E}_\mathbf{p} \cup \mathcal{S}_\mathbf{p})$. Since $f(\mathcal{E}_\mathbf{p} \cup \mathcal{S}_\mathbf{p})$ is a closed set and $\left\{f\left(\mathbf{v}^1\right), \ldots, f\left(\mathbf{v}^{2^m}\right)\right\}$ is a finite set one can choose $\epsilon > 0$ so that the disc of radius ϵ centered at z_0 contains no points of $\partial f(\mathbf{B})$ other than z_0, i.e.,

$$C_\epsilon = \{z \ : \ |z - z_0| < \epsilon\}, \text{ and } C_\epsilon \bigcap \partial f(\mathbf{B}) = z_0.$$

Let C_ϵ^+ be the upper semidisc, and C_ϵ^- be the lower semidisc, i.e.,

$$C_\epsilon^+ = \{z \ : \ |z - z_0| < \epsilon, \text{ Re } z > \text{Re } z_0\}, \text{ and } C_\epsilon^- = \{z \ : \ |z - z_0| < \epsilon, \text{ Re } z < \text{Re } z_0\}.$$

If z_1, $z_2 \in C_\epsilon^+$, such that $z_1 \in f(\mathbf{B})$, and $z_2 \notin f(\mathbf{B})$, then the closed interval $\{\lambda z_1 + (1 - \lambda) z_2\}$, $0 \leq \lambda \leq 1$ must contain a boundary point of $f(\mathbf{B})$ other than

z_0. This shows that either $C_\epsilon^+ \subset f(\mathbf{B})$, or $C_\epsilon^+ \cap f(\mathbf{B}) = \emptyset$. If $C_\epsilon^+ \subset f(\mathbf{B})$, then the uniqueness of the boundary point yields $C_\epsilon^- \subset f(\mathbf{B})$, and $C_\epsilon \subset f(\mathbf{B})$. This show that z_0 may not be a boundary point, and contradicts the assumption. Similar arguments show that the assumption $C_\epsilon^+ \cap f(\mathbf{B}) = \emptyset$ also leads to a contradiction. The proof of Theorem 4.4.1 is now completed.

4.5. Interval Plants with Interval Controllers

In this section we concentrate on the SISO control system given below.

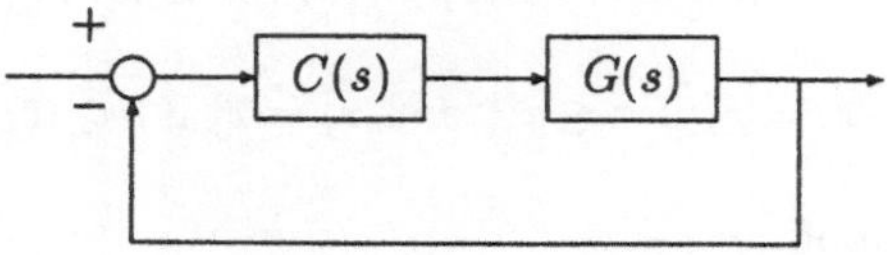

Here

$$C(s) = \frac{\mathbf{p}(s, \mathbf{a}^1)}{\mathbf{p}(s, \mathbf{a}^3)} \text{ and } G(s) = \frac{\mathbf{p}(s, \mathbf{a}^2)}{\mathbf{p}(s, \mathbf{a}^4)}$$

with interval polynomials $\mathbf{p}(s, \mathbf{a}^i) = a_0^i + a_1^i s + \ldots + a_{n_i}^i s^{n_i}$, $i = 1, 2, 3, 4$. The characteristic polynomials of the system are

$$\mathbf{P} = \left\{\mathbf{p}\left(s, \mathbf{a}^1\right)\mathbf{p}\left(s, \mathbf{a}^2\right) + \mathbf{p}\left(s, \mathbf{a}^3\right)\mathbf{p}\left(s, \mathbf{a}^4\right)\right\}, \tag{4.5.1}$$

where $\mathbf{a}^i \in \mathbf{R}^{n_i}$, $\underline{a_{k_i}^i} \le a_{k_i}^i \le \overline{a_{k_i}^i}$, $i = 1, \ldots, 4$, and $k_i = 0, \ldots, n_i$. (Notice that $\mathbf{p}\left(s, \mathbf{a}^i\right)$ define different polynomials for different $\mathbf{a}^i$.)

Remark 4.5.1 Due to the bilinear structure of the perturbations relations (4.2.1) yield a system of linear equations that can be solved analytically.

In order to avoid the "degree dropping" we assume that the leading coefficients of the polynomials in (4.5.1) do not vanish. The set

$$\mathcal{P}_w^1 = \left\{\mathbf{p}(jw, \mathbf{a}^1) \ : \ \underline{a_k^1} \le a_k^1 \le \overline{a_k^1},\ k = 0, \ldots, n_1\right\}$$

is a rectangle in the complex plane with the vertices generated by four Kharitonov polynomials.

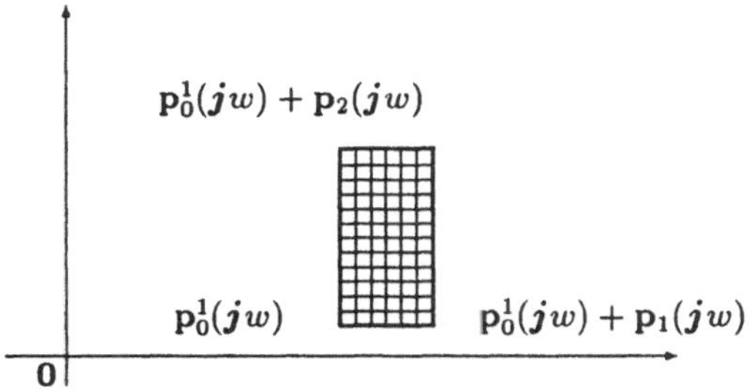

Here

$$\begin{aligned} \mathbf{p}_0^1(s) &= \underline{a_0^1} + \underline{a_1^1}s + \overline{a_2^1}s^2 + \overline{a_3^1}s^3 + \quad \dots \\ \mathbf{p}_1(s) &= [\overline{a_0^1} - \underline{a_0^1}] + [\underline{a_2^1} - \overline{a_2^1}]s^2 + \quad \dots \qquad (4.5.2) \\ \mathbf{p}_2(s) &= [\overline{a_1^1} - \underline{a_1^1}]s + [\underline{a_3^1} - \overline{a_3^1}]s^3 + \quad \dots \end{aligned}$$

The set $\mathcal{P}_w^1$ is the image of the two dimensional box $\{0 \le x_1,\ x_2 \le 1\}$ under the affine transformation

$$(x_1, x_2) \rightarrow \mathbf{p}_0^1(\boldsymbol{j}w) + x_1\mathbf{p}_1(\boldsymbol{j}w) + x_2\mathbf{p}_2(\boldsymbol{j}w),$$

i.e.,

$$\mathcal{P}_w^1 = \left\{\mathbf{p}_0^1(\boldsymbol{j}w) + x_1\mathbf{p}_1(\boldsymbol{j}w) + x_2\mathbf{p}_2(\boldsymbol{j}w)\ :\ 0 \le x_1 \le 1,\ 0 \le x_2 \le 1\right\}. \qquad (4.5.3)$$

Remark 4.5.2 Im $\mathbf{p}_1(\boldsymbol{j}w) = 0$, Re $\mathbf{p}_1(\boldsymbol{j}w) \ge 0$; and Im $\mathbf{p}_2(\boldsymbol{j}w) \ge 0$, Re $\mathbf{p}_2(\boldsymbol{j}w) = 0$. In order to avoid trivial situations we assume that Re $\mathbf{p}_1(\boldsymbol{j}w) > 0$, and Im $\mathbf{p}_2(\boldsymbol{j}w) > 0$ for each $w > 0$.

We define and describe similarly sets $\mathcal{P}_w^2$, $\mathcal{P}_w^3$, and $\mathcal{P}_w^4$. Then the value set of the family (4.5.1) is

$$\mathcal{P}_w = \left\{\mathcal{P}_w^1 \cdot \mathcal{P}_w^2 + \mathcal{P}_w^3 \cdot \mathcal{P}_w^4\right\}, \qquad (4.5.4)$$

where the arithmetic operations are understood in a set–theoretic framework.

Stability of the family (4.5.1) has been investigated by Barmish and Shi [BS1]. The stability criterion proposed in [BS1] requires checking 16 functions of *two* variables w and θ. In contrast to [BS1] we concentrate on principal points. This approach leads to a stability criterion that requires checking stability of one dimensional edges and positivity of a single function that depends on w only.

Let

$$\begin{aligned} F_1(w, \mathbf{x}) &= [\mathbf{p}_0^1(\boldsymbol{j}w) + x_1\mathbf{p}_1(\boldsymbol{j}w) + x_2\mathbf{p}_2(\boldsymbol{j}w)]\,[\mathbf{p}_0^2(\boldsymbol{j}w) + x_3\mathbf{p}_3(\boldsymbol{j}w) + x_4\mathbf{p}_4(\boldsymbol{j}w)] \\ F_2(w, \mathbf{x}) &= [\mathbf{p}_0^3(\boldsymbol{j}w) + x_5\mathbf{p}_5(\boldsymbol{j}w) + x_6\mathbf{p}_6(\boldsymbol{j}w)]\,[\mathbf{p}_0^4(\boldsymbol{j}w) + x_7\mathbf{p}_7(\boldsymbol{j}w) + x_8\mathbf{p}_8(\boldsymbol{j}w)] \end{aligned}$$

For a fixed w the value set $\mathcal{P}_w$ is the image of the box $\mathbf{B}^8$ under the bilinear mapping

$$F(w,\mathbf{x}) \; : \; \mathbf{B}^8 \to \mathbf{C}, \text{ where } F(w,\mathbf{x}) = F_1(w,\mathbf{x}) + F_2(w,\mathbf{x}). \tag{4.5.5}$$

Remark 4.5.3 Relation (4.5.5) shows that stability of the box family (4.5.1) is in fact an *eight* dimensional problem regardless of the degrees of the interval polynomials $\mathbf{p}\left(s,\mathbf{a}^1\right),\ldots,\mathbf{p}\left(s,\mathbf{a}^4\right)$. This useful observation can be traced to [CB] .

Let $\mathcal{E}$ be the set of the one dimensional edges of $\mathbf{B}^8$. Denote by $\mathcal{F}_{ij}^o$ the set of 2^6 open two dimensional faces of $\mathbf{B}^8$

$$\mathcal{F}_{ij}^o = \left\{\mathbf{x} \; : \; \mathbf{x} \in \mathbf{B}^8, \; 0 < x_i, \; x_j < 1; \; x_k = 0 \text{ or } 1, \; k \neq i,j\right\}. \tag{4.5.6}$$

The set of principal points is described next.

Theorem 4.5.1 The polynomial family (4.5.1) is stable if and only if the following union of one and two dimensional polynomial families is stable

$$\mathbf{X_p} = \mathcal{E} \cup \mathcal{F}_{13}^o \cup \mathcal{F}_{14}^o \cup \mathcal{F}_{23}^o \cup \mathcal{F}_{24}^o \cup \mathcal{F}_{57}^o \cup \mathcal{F}_{58}^o \cup \mathcal{F}_{67}^o \cup \mathcal{F}_{68}^o.$$

Application of the frequency domain approach leads to further reduction of computational effort associated with the robust stability problem. For all but finitely many frequencies w, and for each $\mathcal{F}_{ij}^o$ associated with $\mathbf{X_p}$ there exists *at most* sixteen linear segments $\mathbf{S}_{ij}(w,1)$, ..., $\mathbf{S}_{ij}(w,16)$ that belong to $\mathcal{F}_{ij}^o$ and contain principal points of $\mathcal{F}_{ij}^o$. A linear segment is given by

$$\mathbf{S}_{ij}(w,k) = \{\mathbf{x} \; : \; x_j = c_{ij}(w,k)x_i + d_{ij}(w,k), \; x_l = 0, \text{ or } x_l = 1, \; l \neq i,j\}. \tag{4.5.7}$$

For example

$$\mathbf{S}_{13}(w,1) = \left\{\mathbf{x} \; : \; \mathbf{x} \in \mathbf{B}^8, \; x_3 = c_{13}(w,1)x_1 + d_{13}(w,1), \; x_2 = x_4 = x_5 = \ldots = x_8 = 0\right\},$$

where

$$c_{13}(w,1) = \frac{\mathbf{p}_1(jw)}{\mathbf{p}_3(jw)}\frac{\operatorname{Im}\mathbf{p}_0^2(jw)}{\operatorname{Im}\mathbf{p}_0^1(jw)}, \text{ and } d_{13}(w,1) = \frac{\operatorname{Im}\mathbf{p}_0^2(jw)}{\operatorname{Im}\mathbf{p}_0^1(jw)}\frac{\operatorname{Re}\mathbf{p}_0^1(jw)}{\mathbf{p}_3(jw)} - \frac{\operatorname{Re}\mathbf{p}_0^2(jw)}{\mathbf{p}_3(jw)}.$$

Let $\mathbf{S}_{ij}(w) = \bigcup_{k=1}^{16} \mathbf{S}_{ij}(w,k)$ and

$$\mathbf{S}(w) = \mathbf{S}_{13}(w) \cup \mathbf{S}_{14}(w) \cup \mathbf{S}_{23}(w) \cup \mathbf{S}_{24}(w) \cup \mathbf{S}_{57}(w) \cup \mathbf{S}_{58}(w) \cup \mathbf{S}_{67}(w) \cup \mathbf{S}_{68}(w).$$

Then

$$\mathbf{X_p}(w) \subset \mathcal{E} \bigcup \mathbf{S}(w). \tag{4.5.8}$$

Remark 4.5.4 Generically $\mathbf{S}(w)$ contains at most 32 segments (see Subsection 4.5.2, page 90). For a wide range of frequencies $\mathbf{S}(w) = \emptyset$ (see Section 4.6).

The frequency domain stability criterion is given next.

Theorem 4.5.2 Assume that the one dimensional edges $\mathcal{E}$ are stable. The polynomial family (4.5.1) is stable if and only if

$$0 \notin \{\mathbf{p}(jw, \mathbf{S}(w))\} \text{ for each } w > 0.$$

4.5.1. Eight Faces Theorem

We start to prove Theorem 4.5.1 by showing the following: Let $f(\mathbf{x}) = F(w, \mathbf{x})$, where $F(w, \mathbf{x})$ is a bilinear function given by (4.5.5). For each $z \in \partial f(\mathbf{B}^8)$ there exists a principal point $\mathbf{x}$ such that

$$f(\mathbf{x}) = z, \text{ and } |I_f(\mathbf{x})| \leq 2. \tag{4.5.9}$$

First consider a bilinear function

$$f(\mathbf{x}) = f(x_1, x_2, x_3, x_4) = \left[t_0^1 + t_1 x_1 + t_2 x_2\right] \cdot \left[t_0^2 + t_3 x_3 + t_4 x_4\right] + t_0^0,$$

where t stands for a complex number $a + jb$.

Lemma 4.5.1 *Let* $\mathbf{x} \in \mathbf{B}^4$, *be such that* $f(\mathbf{x}) \in \partial f(\mathbf{B}^4)$ *and*

$$\forall \mathbf{y} \in \mathbf{B}^4 \text{ with } f(\mathbf{y}) = f(\mathbf{x}) \text{ one has } |I_f(\mathbf{x})| \leq |I_f(\mathbf{y})|. \tag{4.5.10}$$

Then $|I_f(\mathbf{x})| \leq 2$.

Proof: If both x_1 and x_2 are free coordinates, then due to Lemma 4.2.1

$$t_1\left[t_0^2 + t_3 x_3 + t_4 x_4\right] = G_1(\mathbf{x}) = cG_2(\mathbf{x}) = ct_2\left[t_0^2 + t_3 x_3 + t_4 x_4\right], \quad \operatorname{Im} c = 0.$$

This condition yields

$$f(\mathbf{x}(\gamma)) = f(\mathbf{x}) \text{ for each real } \gamma, \text{ and } \mathbf{x}(\gamma) = (x_1 + \gamma, x_2 - c\gamma, x_3, x_4)^t.$$

For some γ_* the point $(x_1 + \gamma_*, x_2 - c\gamma_*)$ belongs to the boundary of the box $\mathbf{B}^2$, i.e.,

$$f(\mathbf{x}(\gamma_*)) = f(\mathbf{x}) \text{ and } |I_f(\mathbf{x}(\gamma_*))| < |I_f(\mathbf{x})|.$$

This contradicts (4.5.10) and shows that either x_1, or x_2 is an extremal coordinate. For the same reason either x_3, or x_4 is an extremal coordinate. This completes the proof.

Next consider a bilinear function f defined in $\mathbf{B}^8$ by $f(\mathbf{x}) = f_1(\mathbf{x}^1) + f_2(\mathbf{x}^2)$, where

$$\begin{aligned} \mathbf{x}^1 = (x_1, x_2, x_3, x_4) \quad \text{and} \quad f_1(\mathbf{x}^1) &= [t_0^1 + t_1 x_1 + t_2 x_2]\,[t_0^2 + t_3 x_3 + t_4 x_4]\,, \\ \mathbf{x}^2 = (x_5, x_6, x_7, x_8) \quad \text{and} \quad f_2(\mathbf{x}^2) &= [t_0^3 + t_5 x_5 + t_6 x_6]\,[t_0^4 + t_7 x_7 + t_8 x_8]\,. \end{aligned} \tag{4.5.11}$$

Lemma 4.5.2 *Let* $\mathbf{x} \in \mathbf{B}^8$, *be such that* $f(\mathbf{x}) \in \partial f(\mathbf{B}^8)$ *and*

$$\forall \mathbf{y} \in \mathbf{B}^8 \text{ with } f(\mathbf{y}) = f(\mathbf{x}) \text{ one has } |I_f(\mathbf{x})| \le |I_f(\mathbf{y})|. \tag{4.5.12}$$

Then $|I_f(\mathbf{x})| \le 2$.

Proof: Due to Lemma 4.5.10 each $\mathbf{x}^i$, $i = 1, 2$ contains at most 2 free coordinates. We next show that either $\mathbf{x}^1$, or $\mathbf{x}^2$ contains no free coordinates. Suppose the opposite, i.e., there exist free coordinates x_i, $1 \le i \le 4$, and x_k, $5 \le k \le 8$. To simplify the exposition we assume that $i = 1$, and $k = 5$. Due to Lemma 4.2.1

$$G_1(\mathbf{x}) = cG_5(\mathbf{x}), \qquad \operatorname{Im} c = 0.$$

Let $\mathbf{x}(\gamma) = (x_1 + \gamma, x_2, \ldots, x_4, x_5 - \gamma c, x_6, \ldots, x_8)^t$. Then

$$f(\mathbf{x}(\gamma)) = f_1(\mathbf{x}^1) + \gamma G_1(\mathbf{x}) + f_2(\mathbf{x}^2) - \gamma c G_5(\mathbf{x}) = f_1(\mathbf{x}^1) + f_2(\mathbf{x}^2) = f(\mathbf{x}).$$

For some γ_* the point $(x_1 + \gamma_*, x_5 - c\gamma_*)$ hits the boundary of the box $\mathbf{B}^2$, i.e.,

$$f(\mathbf{x}(\gamma_*)) = f(\mathbf{x}), \text{ and } |I_f(\mathbf{x}(\gamma_*))| < |I_f(\mathbf{x})|.$$

This contradiction completes the proof of the lemma.

To complete the proof of (4.5.9) we notice that each $\mathbf{x} \in \mathbf{B}^8$ with $f(\mathbf{x}) \in \partial f(\mathbf{B}^8)$ is a principal point.

In the remainder of the section we select open two dimensional faces of $\mathbf{B}^8$ that contain principal points and complete the proof of Theorem 4.5.1.

Let $\mathbf{x}$ be a principal point that belongs to $\mathcal{F}_{ik}^o$. The partial derivatives $G_i(w, \mathbf{x})$ of $F(w, \mathbf{x})$ are given next

$$\begin{aligned} G_1 &= \mathbf{p}_1(jw)\,[\mathbf{p}_0^2(jw) + x_3\mathbf{p}_3(jw) + x_4\mathbf{p}_4(jw)] \\ G_2 &= \mathbf{p}_2(jw)\,[\mathbf{p}_0^2(jw) + x_3\mathbf{p}_3(jw) + x_4\mathbf{p}_4(jw)] \\ G_3 &= \mathbf{p}_3(jw)\,[\mathbf{p}_0^1(jw) + x_1\mathbf{p}_1(jw) + x_2\mathbf{p}_2(jw)] \\ G_4 &= \mathbf{p}_4(jw)\,[\mathbf{p}_0^1(jw) + x_1\mathbf{p}_1(jw) + x_2\mathbf{p}_2(jw)] \\ G_5 &= \mathbf{p}_5(jw)\,[\mathbf{p}_0^4(jw) + x_7\mathbf{p}_7(jw) + x_8\mathbf{p}_8(jw)] \\ G_6 &= \mathbf{p}_6(jw)\,[\mathbf{p}_0^4(jw) + x_7\mathbf{p}_7(jw) + x_8\mathbf{p}_8(jw)] \\ G_7 &= \mathbf{p}_7(jw)\,[\mathbf{p}_0^3(jw) + x_5\mathbf{p}_5(jw) + x_6\mathbf{p}_6(jw)] \\ G_8 &= \mathbf{p}_8(jw)\,[\mathbf{p}_0^3(jw) + x_5\mathbf{p}_5(jw) + x_6\mathbf{p}_6(jw)] \end{aligned}$$

According to Remark 4.5.2 $G_1(w,\mathbf{x}) \asymp G_3(w,\mathbf{x})$ if and only if

$$\mathbf{p}_0^2(jw) + x_3\mathbf{p}_3(jw) + x_4\mathbf{p}_4(jw) \asymp \mathbf{p}_0^1(jw) + x_1\mathbf{p}_1(jw) + x_2\mathbf{p}_2(jw).$$

For computational convenience for each frequency w we introduce four complex valued functions of two real variables γ_1 and γ_2.

$$\begin{aligned}
z_1(\gamma_1,\gamma_2) &= \mathbf{p}_0^1(jw) + \gamma_1\mathbf{p}_1(jw) + \gamma_2\mathbf{p}_2(jw),\\
z_2(\gamma_1,\gamma_2) &= \mathbf{p}_0^2(jw) + \gamma_1\mathbf{p}_3(jw) + \gamma_2\mathbf{p}_4(jw),\\
z_3(\gamma_1,\gamma_2) &= \mathbf{p}_0^3(jw) + \gamma_1\mathbf{p}_5(jw) + \gamma_2\mathbf{p}_6(jw),\\
z_4(\gamma_1,\gamma_2) &= \mathbf{p}_0^4(jw) + \gamma_1\mathbf{p}_7(jw) + \gamma_2\mathbf{p}_8(jw).
\end{aligned}$$

The partial derivatives $G_i(w,\mathbf{x})$ satisfy

$$\begin{array}{llll}
G_1 \asymp z_2(x_3,x_4), & G_3 \asymp z_1(x_1,x_2), & G_5 \asymp z_4(x_7,x_8), & G_7 \asymp z_3(x_5,x_6),\\
G_2 \asymp jz_2(x_3,x_4), & G_4 \asymp jz_1(x_1,x_2), & G_6 \asymp jz_4(x_7,x_8), & G_8 \asymp jz_3(x_5,x_6).
\end{array} \tag{4.5.13}$$

To simplify the exposition we set $i = 1$, and consider in detail seven cases $k = 2,\dots 8$.

Case 2. $0 < x_1 < 1$, $0 < x_2 < 1$. The existence of $\mathbf{x}$ yields $z_2(e_3,e_4) \asymp jz_2(e_3,e_4)$, and shows that $\mathcal{F}_{12}^o$ contains no principal points. (Here, for example, e_3 stands for an extreme point of the interval $[0,1]$, i.e., $e_3 = 0$, or $e_3 = 1$.)

Case 3. $0 < x_1 < 1$, $0 < x_3 < 1$. In this case $z_2(x_3,e_4) \asymp z_1(x_1,e_2)$, i.e.,

$$\begin{aligned}
x_3 &= x_1 c\frac{\mathbf{p}_1(jw)}{\mathbf{p}_3(jw)} + \left[c\frac{\text{Re}\,\mathbf{p}_0^1(jw)}{\mathbf{p}_3(jw)} - \frac{\text{Re}\,\mathbf{p}_0^2(jw)}{\mathbf{p}_3(jw)}\right]\\
0 &= \left[\text{Im}\ \mathbf{p}_0^2(jw) + e_4\text{Im}\ \mathbf{p}_4(jw)\right] - c\left[\text{Im}\ \mathbf{p}_0^1(jw) + e_2\text{Im}\ \mathbf{p}_2(jw)\right].
\end{aligned} \tag{4.5.14}$$

There exists a finite set of frequencies so that

$$\text{Im}\ \mathbf{p}_0^1(jw) + e_2\text{Im}\ \mathbf{p}_2(jw) = 0, \text{ and Im}\ \mathbf{p}_0^2(jw) + e_4\text{Im}\ \mathbf{p}_4(jw) = 0. \tag{4.5.15}$$

If (4.5.15) does not hold, then

$$c = c(w,e_2,e_4) = \frac{\text{Im}\ \mathbf{p}_0^2(jw) + e_4\text{Im}\ \mathbf{p}_4(jw)}{\text{Im}\ \mathbf{p}_0^1(jw) + e_2\text{Im}\ \mathbf{p}_2(jw)},$$

and solutions of the system (4.5.14) form at most four straight lines in the (x_1,x_3) plane.

Case 4. $0 < x_1 < 1$, $0 < x_4 < 1$. In this case $z_2(e_3,x_4) \asymp jz_1(x_1,e_2)$. The corresponding system of linear equations with respect to x_1 and x_4 is

$$\begin{aligned}
0 &= \left[\text{Re}\ \mathbf{p}_0^2(jw) + e_3\mathbf{p}_3(jw)\right] + c\left[\text{Im}\ \mathbf{p}_0^1(jw) + e_2\text{Im}\ \mathbf{p}_2(jw)\right]\\
x_4 &= x_1 c\frac{\mathbf{p}_1(jw)}{\text{Im}\,\mathbf{p}_4(jw)} + \left[c\frac{\text{Re}\,\mathbf{p}_0^1(jw)}{\text{Im}\,\mathbf{p}_4(jw)} - \frac{\text{Re}\,\mathbf{p}_0^2(jw)}{\text{Im}\,\mathbf{p}_4(jw)}\right].
\end{aligned} \tag{4.5.16}$$

Similarly to the previous case solutions of (4.5.16) form at most four straight lines.

Case 5. $0 < x_1 < 1$, $0 < x_5 < 1$. In this case $z_2(e_3, x_4) \asymp z_4(e_7, e_8)$, i.e., $F(w, \mathcal{F}_{15}^o)$ is covered by images of the one dimensional edges

$$\{0 \le x_1 \le 1,\ x_5 = 0,\ x_l = e_l,\ l \ne 1,5\} \text{ and } \{0 \le x_5 \le 1,\ x_1 = 1,\ x_l = e_l,\ l \ne 1,5\}.$$

This case therefore can be disregarded. For the same reason we disregard the sets $\mathcal{F}_{16}^o$, $\mathcal{F}_{17}^o$ and $\mathcal{F}_{18}^o$. Similar arguments applied to the cases $i = 2, \ldots, 7$ complete the proof of Theorem 4.5.1.

4.5.2. Frequency Domain Criterion

In this subsection we describe principal segments $\mathbf{S}(w)$. We start with a simple observation concerning arguments of complex numbers. Let $a + jb$ be a complex number with $a \ne 0$, and $b \ne 0$. For each pair of real $\epsilon_1 < \epsilon_2$ the following relations hold:

$$\begin{aligned} (a+\epsilon_2) + jb &\prec (a+\epsilon_1) + jb \quad \text{if } b > 0 \\ (a+\epsilon_2) + jb &\succ (a+\epsilon_1) + jb \quad \text{if } b < 0 \\ a + j(b+\epsilon_2) &\prec a + j(b+\epsilon_1) \quad \text{if } a > 0 \\ a + j(b+\epsilon_2) &\succ a + j(b+\epsilon_1) \quad \text{if } a < 0. \end{aligned} \tag{4.5.17}$$

In what follows we provide a detailed description of the set $\mathbf{S}_{13}(w)$.

Let $\mathbf{x} \in \mathcal{F}_{13}^o$ be a principal point, then $z_2(x_3, e_4) \asymp G_1(w, \mathbf{x}) \asymp G_3(w, \mathbf{x}) \asymp z_1(x_1, e_2)$, and $z_2(x_3, e_4) = cz_1(x_1, e_2)$. We will assume that Im $z_1(x_1, e_2) \ne 0$, and Im $z_2(x_3, e_4) \ne 0$. The assumptions exclude a finite set of frequencies only. First consider the case $c > 0$. Then either

$$z_1(x_1, e_2),\ z_2(x_3, e_4) \in \mathbf{I} \cup \mathbf{II}, \quad \text{or } z_1(x_1, e_2),\ z_2(x_3, e_4) \in \mathbf{III} \cup \mathbf{IV},$$

where **I, II, III, IV** are the first, second, third and fourth quadrants of the complex plane. For an extremal coordinate e_i satisfying

$$G_i(w, \mathbf{x}) \ne 0, \text{ and } G_i(w, \mathbf{x}) \not\asymp G_1(w, \mathbf{x}) \tag{4.5.18}$$

a straightforward application of (4.5.17) and Lemma 4.4.3 yields the following

$$\begin{aligned} &\text{if } \left\{ \begin{array}{c} c > 0 \\ z_1(x_1, e_2) \in \mathbf{I} \cup \mathbf{II} \\ z_2(x_3, e_4) \in \mathbf{I} \cup \mathbf{II} \end{array} \right\}, \quad \text{then } (-1)^{e_i} G_i(w, \mathbf{x}) \succ G_1(w, \mathbf{x}) \\ &\text{if } \left\{ \begin{array}{c} c > 0 \\ z_1(x_1, e_2) \in \mathbf{III} \cup \mathbf{IV} \\ z_2(x_3, e_4) \in \mathbf{III} \cup \mathbf{IV} \end{array} \right\}, \quad \text{then } (-1)^{e_i} G_i(w, \mathbf{x}) \prec G_1(w, \mathbf{x}). \end{aligned} \tag{4.5.19}$$

Analogously, when $c < 0$ one has

$$z_1(x_1, e_2) \in \mathbf{I} \cup \mathbf{II} \text{ and } z_2(x_3, e_4) \in \mathbf{III} \cup \mathbf{IV},$$

or

$$z_1(x_1, e_2) \in \mathbf{III} \cup \mathbf{IV} \text{ and } z_2(x_3, e_4) \in \mathbf{I} \cup \mathbf{II},$$

and when condition (4.5.18) holds one has

$$\begin{aligned} &\text{if } \left\{ \begin{matrix} c < 0 \\ z_1(x_1, e_2) \in \mathbf{I} \cup \mathbf{II} \\ z_2(x_3, e_4) \in \mathbf{III} \cup \mathbf{IV} \end{matrix} \right\}, \quad \text{then} \quad (-1)^{e_i} G_i(w, \mathbf{x}) \prec G_1(w, \mathbf{x}) \\ &\text{if } \left\{ \begin{matrix} c < 0 \\ z_1(x_1, e_2) \in \mathbf{III} \cup \mathbf{IV} \\ z_2(x_3, e_4) \in \mathbf{I} \cup \mathbf{II} \end{matrix} \right\}, \quad \text{then} \quad (-1)^{e_i} G_i(w, \mathbf{x}) \succ G_1(w, \mathbf{x}). \end{aligned} \tag{4.5.20}$$

Hence the extremal coordinate e_i satisfying (4.5.18) is uniquely determined by relations (4.5.19) and (4.5.20). Note that

$$\begin{aligned} &G_2(w, \mathbf{x}) \neq 0, \ G_4(w, \mathbf{x}) \neq 0, \\ &\arg G_2(w, \mathbf{x}) = \arg G_1(w, \mathbf{x}) + \tfrac{\pi}{2}, \\ &\arg G_4(w, \mathbf{x}) = \arg G_3(w, \mathbf{x}) + \tfrac{\pi}{2}. \end{aligned}$$

This immediately implies

$$\begin{aligned} &e_2 = 0, \quad e_4 = 0 \quad \text{if} \quad c > 0, \quad z_1(x_1, e_2) \in \mathbf{I} \cup \mathbf{II}, \quad z_2(x_3, e_4) \in \mathbf{I} \cup \mathbf{II}. \\ &e_2 = 1, \quad e_4 = 1 \quad \text{if} \quad c > 0, \quad z_1(x_1, e_2) \in \mathbf{III} \cup \mathbf{IV}, \quad z_2(x_3, e_4) \in \mathbf{III} \cup \mathbf{IV}. \\ &e_2 = 1, \quad e_4 = 0 \quad \text{if} \quad c < 0, \quad z_1(x_1, e_2) \in \mathbf{I} \cup \mathbf{II}, \quad z_2(x_3, e_4) \in \mathbf{III} \cup \mathbf{IV}. \\ &e_2 = 0, \quad e_4 = 1 \quad \text{if} \quad c < 0, \quad z_1(x_1, e_2) \in \mathbf{III} \cup \mathbf{IV}, \quad z_2(x_3, e_4) \in \mathbf{I} \cup \mathbf{II}. \end{aligned}$$

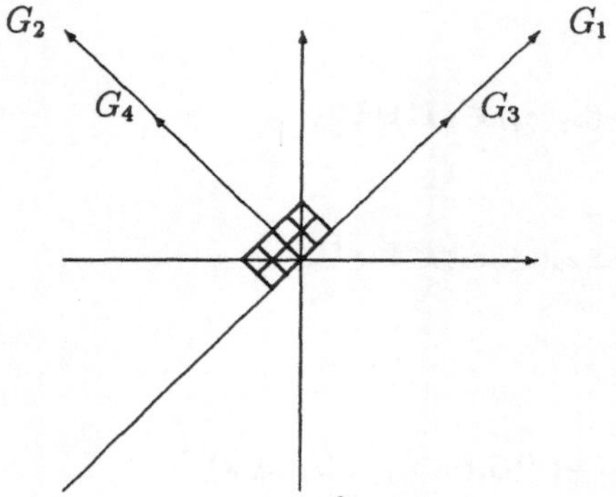

$c > 0$
$z_1(x_1, e_2) \in \mathbf{I} \cup \mathbf{II}$
$z_2(x_3, e_4) \in \mathbf{I} \cup \mathbf{II}$
$e_2 = 0,\ e_4 = 0$

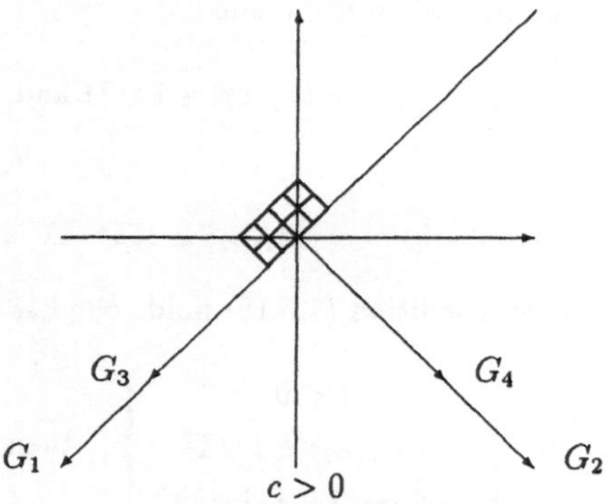

$c > 0$
$z_1(x_1, e_2) \in \mathbf{III} \cup \mathbf{IV}$
$z_2(x_3, e_4) \in \mathbf{III} \cup \mathbf{IV}$
$e_2 = 1,\ e_4 = 1$

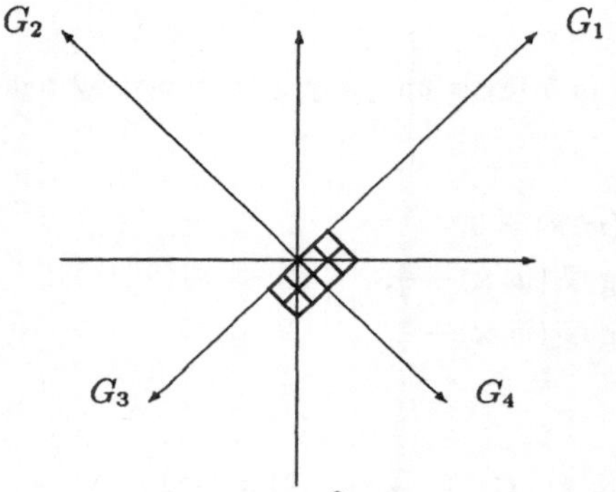

$c < 0$
$z_1(x_1, e_2) \in \mathbf{I} \cup \mathbf{II}$
$z_2(x_3, e_4) \in \mathbf{III} \cup \mathbf{IV}$
$e_2 = 1,\ e_4 = 0$

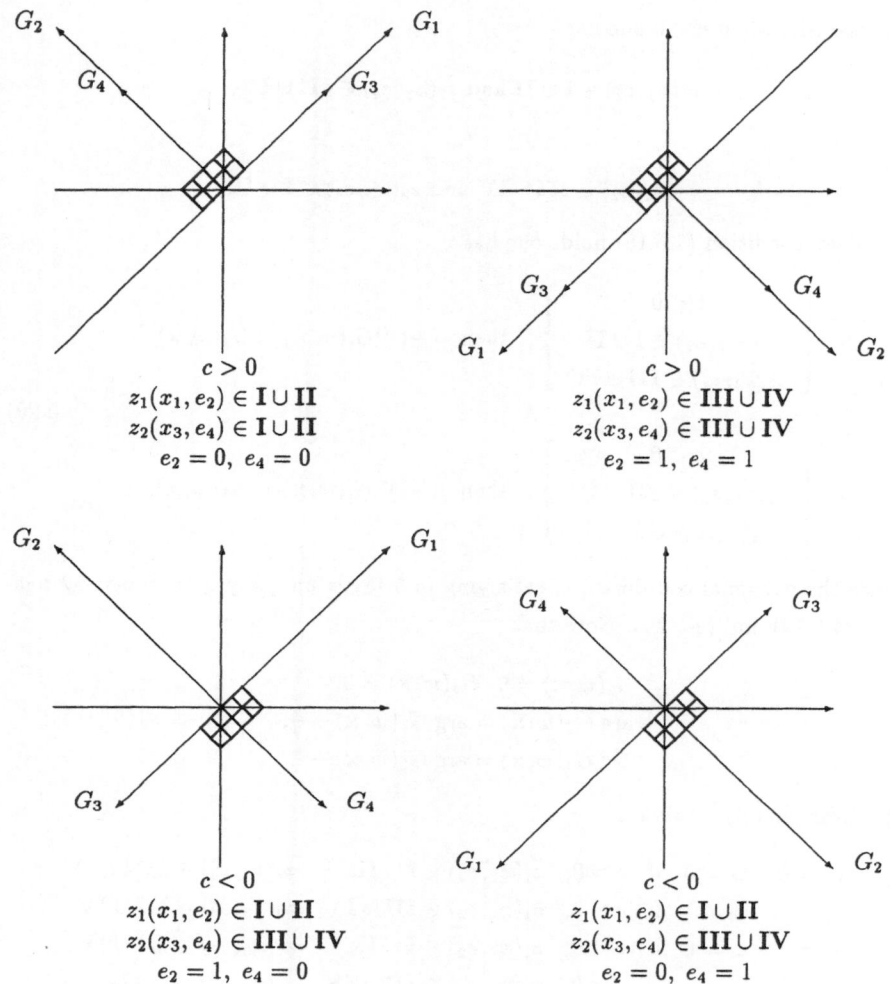

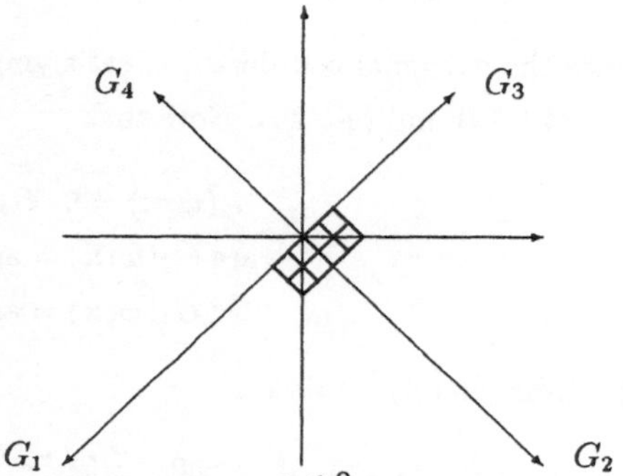

$c < 0$
$z_1(x_1, e_2) \in \mathbf{I} \cup \mathbf{II}$
$z_2(x_3, e_4) \in \mathbf{III} \cup \mathbf{IV}$
$e_2 = 0,\ e_4 = 1$

On the other hand

$$\{G_5(w, \mathbf{x}),\ G_6(w, \mathbf{x})\} \text{ and } \{G_7(w, \mathbf{x}),\ G_8(w, \mathbf{x})\} \tag{4.5.21}$$

are pairs of mutually orthogonal two dimensional vectors. So if, for example,

$$G_5(w, \mathbf{x}) \asymp G_1(w, \mathbf{x})$$

and $x_5 = 0$, or $x_5 = 1$, the value of x_6 is uniquely determined by (4.5.19) and (4.5.20). This shows that *at most* 2 extremal coordinates may not be determined uniquely by

(4.5.19) and (4.5.20). Hence there exist *at most* sixteen open two dimensional faces $\mathbf{F}^o_{13}(w,k)$, $k = 1,\dots,16$ that contain principal points of $\mathcal{F}^o_{13}$. We denote principal segments that belong to the two dimensional faces by $\mathbf{S}_{13}(w,k)$, $k = 1,\dots,16$ respectively, and denote principal points of $\mathcal{F}^o_{13}$ at w by $\mathbf{S}_{13}(w)$, i.e., $\mathbf{S}_{13}(w) = \bigcup_{k=1}^{16} \mathbf{S}_{13}(w,k)$.

Remark 4.5.5 Generically $G_5(w,\mathbf{x}) \not\prec G_1(w,\mathbf{x})$, $G_6(w,\mathbf{x}) \not\prec G_1(w,\mathbf{x})$, and $\mathbf{S}_{13}(w)$ contains at most four nonempty segments. For many frequencies the number of segments in $\mathbf{S}_{13}(w)$ reduces to one (see Section 4.6).

Repetition of these arguments shows that at all but finitely many frequencies w one has to check the images of the one dimensional edges $\mathcal{E}$ and the set of *at most* 128 one dimensional segments $\mathbf{S}(w) = \mathbf{S}_{13}(w) \cup \mathbf{S}_{14}(w) \cup \mathbf{S}_{23}(w) \cup \mathbf{S}_{24}(w) \cup \mathbf{S}_{57}(w) \cup \mathbf{S}_{58}(w) \cup \mathbf{S}_{67}(w) \cup \mathbf{S}_{68}(w)$.

4.6. Numerical Example

In this section we consider the polynomial family

$$\mathbf{P} = \left\{\mathbf{p}\left(s,\mathbf{a}^1\right)\mathbf{p}\left(s,\mathbf{a}^2\right) + \mathbf{p}\left(s,\mathbf{a}^3\right)\mathbf{p}\left(s,\mathbf{a}^4\right)\right\} \tag{4.6.1}$$

given in [BS1]. The interval polynomials $\mathbf{p}\left(s,\mathbf{a}^1\right)$, $\mathbf{p}\left(s,\mathbf{a}^2\right)$, $\mathbf{p}\left(s,\mathbf{a}^3\right)$, $\mathbf{p}\left(s,\mathbf{a}^4\right)$ are defined by

$$\mathbf{p}(s,\mathbf{a}^1) = a_0^1 + a_1^1 s, \quad \mathbf{p}(s,\mathbf{a}^3) = a_0^3 + a_1^3 s + s^2,$$
$$\mathbf{p}(s,\mathbf{a}^2) = a_0^2 + a_1^2 s, \quad \mathbf{p}(s,\mathbf{a}^4) = a_0^4 + a_1^4 s + s^2,$$

and two sets of interval parameters:

Set 1.

$$\begin{array}{llll}
\underline{a_0^1} = 1.7, & \overline{a_0^1} = 2.3; & \underline{a_1^1} = 2.7, & \overline{a_1^1} = 3.3 \\
\underline{a_0^2} = 22.9, & \overline{a_0^2} = 23.1; & \underline{a_1^2} = 19.9, & \overline{a_1^2} = 20.1 \\
\underline{a_0^3} = 9.5, & \overline{a_0^3} = 10.5; & \underline{a_1^3} = -3.5, & \overline{a_1^3} = -2.5 \\
\underline{a_0^4} = 4.9, & \overline{a_0^4} = 5.1; & \underline{a_1^4} = 9.9, & \overline{a_1^4} = 10.1
\end{array}$$

Set 2.

$$\begin{array}{llll}
\underline{a_0^1} = 1.7, & \overline{a_0^1} = 2.3; & \underline{a_1^1} = 2.7, & \overline{a_1^1} = 3.3 \\
\underline{a_0^2} = 22.8, & \overline{a_0^2} = 23.2; & \underline{a_1^2} = 19.8, & \overline{a_1^2} = 20.2 \\
\underline{a_0^3} = 9.5, & \overline{a_0^3} = 10.5; & \underline{a_1^3} = -3.5, & \overline{a_1^3} = -2.5 \\
\underline{a_0^4} = 4.8, & \overline{a_0^4} = 5.2; & \underline{a_1^4} = 9.8, & \overline{a_1^4} = 10.2
\end{array}$$

A straightforward computation shows that in both cases conditions of Theorem 4.3.2 and Remark 4.5.2 are met. To investigate stability of the family (4.6.1) one can apply Theorem 4.5.2.

In what follows we identify the set of principal segments $\mathbf{S}(w)$ at a given frequency w, and investigate stability of the family (4.6.1). The symbolic manipulations necessary for this investigation have been performed using MAPLE V, while the corresponding numerical computations have been performed using MATLAB.

When $G_1 \asymp G_3$ one has $z_2(x_3, e_4) \asymp z_1(x_1, e_2)$ (see page 89). A straightforward computation shows that for each $w > 0$

$$\min_{x_1,e_2,x_3,e_4} \begin{vmatrix} \text{Im } z_1(x_1,e_2) & \text{Im } z_2(x_3,e_4) \\ \text{Re } z_1(x_1,e_2) & \text{Re } z_2(x_3,e_4) \end{vmatrix} \geq \min_{x_1,x_2,x_3,x_4} \begin{vmatrix} \text{Im } z_1(x_1,x_2) & \text{Im } z_2(x_3,x_4) \\ \text{Re } z_1(x_1,x_2) & \text{Re } z_2(x_3,x_4) \end{vmatrix} =$$

$$w \min_{x_1,x_2,x_3,x_4} [27.9 + 1.08x_3 + 13.68x_2 + 0.24x_2x_3 - 0.68x_4 - 11.88x_1 - 0.24x_1x_4] > 0.$$

Hence the two dimensional faces $\mathcal{F}_{13}^o$ contain no principal points. For the same reason the two dimensional faces $\mathcal{F}_{14}^o$, $\mathcal{F}_{23}^o$, and $\mathcal{F}_{24}^o$ contain no principal points. We focus now on the two dimensional faces $\mathcal{F}_{57}^o$, $\mathcal{F}_{58}^o$, $\mathcal{F}_{67}^o$, and $\mathcal{F}_{68}^o$.

Case 1: $\mathbf{x} \in \mathcal{F}_{57}^o$ and $G_5 \asymp G_7$, where

$$G_5 \asymp z_4(x_7, e_8) = \left[\underline{a_0^4} + x_7\left(\overline{a_0^4} - \underline{a_0^4}\right) - w^2\right] + jw\left[\underline{a_1^4} + e_8\left(\overline{a_1^4} - \underline{a_1^4}\right)\right],$$

$$G_7 \asymp z_3(x_5, e_6) = \left[\underline{a_0^3} + x_5\left(\overline{a_0^3} - \underline{a_0^3}\right) - w^2\right] + jw\left[\underline{a_1^3} + e_6\left(\overline{a_1^3} - \underline{a_1^3}\right)\right].$$

As w moves from 0 to ∞ arg $G_5(w, \mathbf{x})$ increases from 0 to π, and arg $G_7(w, \mathbf{x})$ decreases from 0 to $-\pi$. The equation $G_5(w, \mathbf{x}) \asymp G_7(w, \mathbf{x})$ has a unique solution w_1 so that

$$\frac{\pi}{2} < \arg G_5(w_1, \mathbf{x}) = \pi + \arg G_7(w_1, \mathbf{x}) < \pi.$$

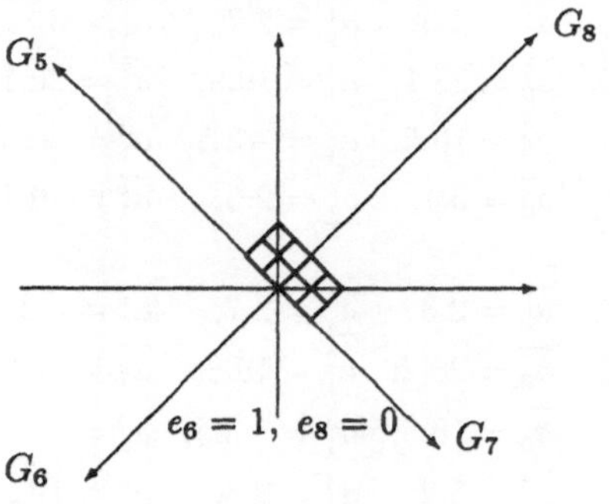

An application of Lemma 4.4.3 yields $e_6 = 1$, and $e_8 = 0$. The relation $z_4(x_7, 0) \asymp$

$z_3(x_5,1)$ yields the following quadratic interval equation with respect to w

$$\frac{\underline{a_0^4}+x_7\left(\overline{a_0^4}-\underline{a_0^4}\right)-w^2}{\underline{a_0^3}+x_5\left(\overline{a_0^3}-\underline{a_0^3}\right)-w^2}=\frac{\underline{a_1^4}}{\overline{a_1^3}}.$$

A straightforward computation shows that the range for w is $w \in [2.9231,\ 3.0696]$, and

$$x_7=\frac{\underline{a_1^4}\left(\overline{a_0^3}-\underline{a_0^3}\right)}{\overline{a_1^3}\left(\overline{a_0^4}-\underline{a_0^4}\right)}x_5+\frac{\underline{a_1^4a_0^3}-\underline{a_1^3a_0^4}+\left(\underline{a_1^3}-\underline{a_1^4}\right)w^2}{\overline{a_1^3}\left(\overline{a_0^4}-\underline{a_0^4}\right)}. \tag{4.6.2}$$

Furthermore, $G_1 \prec G_5$, $G_3 \prec G_5$, $G_5 \prec G_2$, and $G_5 \prec G_4$. These relations yield

$$e_1=0,\ e_2=1,\ e_3=0,\ e_4=1,\ e_6=1,\ e_8=0.$$

The analysis associated with the faces $\mathcal{F}_{58}^o$, $\mathcal{F}_{67}^o$, and $\mathcal{F}_{68}^o$ is similar to that presented above and will be omitted. The summary of results related to principal points of those faces is given next.

Case 2: $\mathbf{x} \in \mathcal{F}_{58}^o$, $e_6=1,\ e_7=0$, or $e_6=1,\ e_7=1$.

If $e_6=1,\ e_7=0$, then $w \in [1.0724,\ 1.1600]$, and

$$x_8=\frac{\left(\overline{a_0^3}-\underline{a_0^3}\right)\left(\underline{a_0^4}-w^2\right)}{-w^2\overline{a_1^3}\left(\overline{a_1^4}-\underline{a_1^4}\right)}x_5+\frac{\underline{a_0^3a_0^4}+\left(\underline{a_1^4}\overline{a_1^3}-\underline{a_0^3}-\underline{a_0^4}\right)w^2+w^4}{-w^2\overline{a_1^3}\left(\overline{a_1^4}-\underline{a_1^4}\right)}. \tag{4.6.3}$$

If $e_6=1,\ e_7=1$, then $w \in [6.1444,\ 6.3217]$, and

$$x_8=\frac{\left(\overline{a_0^3}-\underline{a_0^3}\right)\left(\overline{a_0^4}-w^2\right)}{-w^2\overline{a_1^3}\left(\overline{a_1^4}-\underline{a_1^4}\right)}x_5+\frac{\underline{a_0^3}\overline{a_0^4}+\left(\underline{a_1^4}\overline{a_1^3}-\underline{a_0^3}-\overline{a_0^4}\right)w^2+w^4}{-w^2\overline{a_1^3}\left(\overline{a_1^4}-\underline{a_1^4}\right)}. \tag{4.6.4}$$

Case 3: $\mathbf{x} \in \mathcal{F}_{67}^o$, $e_5=0,\ e_8=0$ or $e_5=1,\ e_8=0$.

If $e_5=0,\ e_8=0$, then $w \in [0.9741,\ 1.1512]$, and

$$x_7=\frac{-w^2\underline{a_1^4}\left(\overline{a_1^3}-\underline{a_1^3}\right)}{\left(\underline{a_0^3}-w^2\right)\left(\overline{a_0^4}-\underline{a_0^4}\right)}x_6+\frac{-\underline{a_0^4a_0^3}+\left(\underline{a_0^3}+\underline{a_0^4}-\underline{a_1^4a_1^3}\right)w^2-w^4}{\left(\underline{a_0^3}-w^2\right)\left(\overline{a_0^4}-\underline{a_0^4}\right)}. \tag{4.6.5}$$

If $e_5=1,\ e_8=0$, then $w \in [6.1949,\ 6.9979]$, and

$$x_7=\frac{-w^2\underline{a_1^4}\left(\overline{a_1^3}-\underline{a_1^3}\right)}{\left(\overline{a_0^3}-w^2\right)\left(\overline{a_0^4}-\underline{a_0^4}\right)}x_6+\frac{-\underline{a_0^4}\overline{a_0^3}+\left(\overline{a_0^3}+\underline{a_0^4}-\underline{a_1^4a_1^3}\right)w^2-w^4}{\left(\overline{a_0^3}-w^2\right)\left(\overline{a_0^4}-\underline{a_0^4}\right)}. \tag{4.6.6}$$

Case 4: $\mathbf{x} \in \mathcal{F}_{68}^o$, $e_5=0,\ e_7=1$, $w \in [2.8090,\ 3.0855]$, and

$$x_8=\frac{\left(\overline{a_0^4}-w^2\right)\left(\overline{a_1^3}-\underline{a_1^3}\right)}{\left(\underline{a_0^3}-w^2\right)\left(\overline{a_1^4}-\underline{a_1^4}\right)}x_6+\frac{\underline{a_1^3}\overline{a_0^4}-\underline{a_1^4a_0^3}+\left(\underline{a_1^4}-\underline{a_1^3}\right)w^2}{\left(\underline{a_0^3}-w^2\right)\left(\overline{a_1^4}-\underline{a_1^4}\right)}. \tag{4.6.7}$$

In order to check stability of the polynomial family (4.6.1) one has

- to check stability of all the one dimensional exposed edges of $\mathbf{B}^8$,
- to verify that the images of **six** linear segments

$$\{x_k = cx_i + d \; : \; 0 < x_i, \; x_k < 1, \; x_j = e_j \text{ if } j \neq i, \; k\}$$

at specified frequencies w do not contain the origin.

The description of the frequencies and the segments is given below.

#	frequencies and equations $x_k = cx_i + d$ are given by	values of extremal coordinates
1.	(4.6.2)	$e_1 = 0, \; e_2 = 1, \; e_3 = 0, \; e_4 = 1, \; e_6 = 1, \; e_8 = 0.$
2.	(4.6.3)	$e_1 = 0, \; e_2 = 1, \; e_3 = 0, \; e_4 = 1, \; e_6 = 1, \; e_7 = 0.$
3.	(4.6.4)	$e_1 = 0, \; e_2 = 1, \; e_3 = 0, \; e_4 = 1, \; e_6 = 1, \; e_7 = 1.$
4.	(4.6.5)	$e_1 = 0, \; e_2 = 0, \; e_3 = 0, \; e_4 = 0, \; e_5 = 0, \; e_8 = 0.$
5.	(4.6.6)	$e_1 = 1, \; e_2 = 1, \; e_3 = 1, \; e_4 = 1, \; e_5 = 1, \; e_8 = 0.$
6.	(4.6.7)	$e_1 = 0, \; e_2 = 0, \; e_3 = 0, \; e_4 = 0, \; e_5 = 0, \; e_7 = 1.$

An application of a MATLAB program shows the critical values $w = 5.50$ for the interval **Set 1**, and $w = 5.444$ for the interval **Set 2**. In both cases open two dimensional faces contain no principal segments. The principal points at these frequencies are the one dimensional exposed edges of $\mathbf{B}^8$. The images of the exposed edges at the critical frequencies are given below.

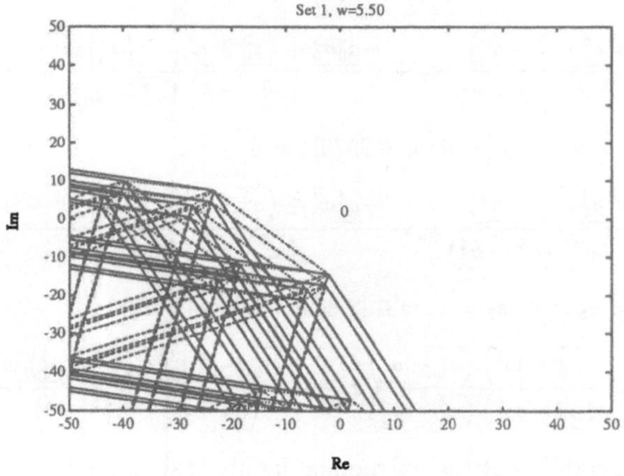

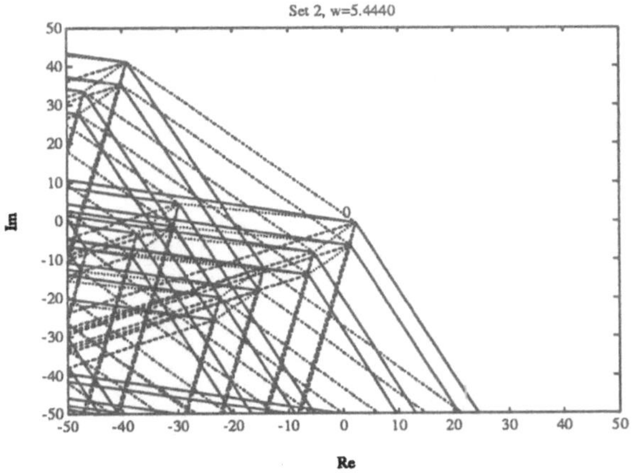

The results obtained show that the polynomial family (4.6.1) is stable with interval constraints of **Set 1**, and unstable with interval constraints of **Set 2**. The results are consistent with those given in [BS1].

The principal points approach leads to computationally tractable robust stability conditions for multiaffine families of transfer functions and quasipolynomial families with interval coefficients and interval delays. For investigations of value sets of multiaffine families of transfer functions we refer the reader to Fu, Dasgupta and Blondel [FDB], and Polyak and Lan [PL]. Interval quasipolynomials are tackled in Chapter 6 of this book.

Chapter 5

Multidimensional Systems and Systems with Commensurate Delays

This chapter is concerned with two important and related classes of systems: linear time invariant time delay systems and multidimensional digital filters (see e.g., [Kam], [CBL]). Stability of both classes is intimately related to stability of the corresponding characteristic polynomials. A brief discussion concerning stability of a linear system and root location of the corresponding characteristic function is provided in the next section. Robust stability of multivariate polynomial and quasipolynomial families is the main issue of the chapter.

The first part of the chapter deals with a family of bivariate polynomials whose coefficients are affine functions of parameters. Following [Ko1] we apply optimization tools and reduce the original robust stability problem to a two dimensional unconstrained global minimization problem. A generalization of this approach to $n-D$ polynomials is straightforward (see e.g., [CM]). However numerical solutions of the corresponding global minimization problems in $\mathbf{R}^n$ are extremely time consuming. Furthermore, a straightforward application of optimization tools to quasipolynomial families with uncertain coefficients and delays leads a family of difficult minimization problems with nonlinear constraints.

It therefore seems reasonable to try alternative approaches. For a general philosophy concerning the "testing family" approach we refer to [Bos3] and [Bos4]. A specific example of the "testing family" approach to robust stability of multivariate polynomial families is given in [SZ]. An illustration of the "testing family" approach to uncertain quasipolynomials is given in Section 5.8. Following [HKZ1] we select a particularly simple testing family for a quasipolynomial family with uncertain coefficients and delays.

For additional references related to the robustness of multivariate systems and related topics we refer the reader to the Special Issue on Robustness of Multidimensional Systems, Multidimensional Systems and Signal Processing, Vol. 5, No. 4, 1994.

5.1. Stability of linear systems and zeros of characteristic polynomials

Stability of a linear time–invariant system arises in many applications. Specific stability requirements are dictated by engineering and physical considerations. One of the most popular types of stability is Bounded–Input–Bounded–Output (BIBO) stability (for detailed discussion of different types of stability including the *practical*–BIBO stability see e.g., [J3]). There is a simple relation between BIBO stability of a 1–D linear system and poles of the corresponding system transfer function. If the numerator and the denominator of the transfer function are relatively prime, then the system is BIBO stable if and only if the transfer function is devoid of poles in a prescribed *closed* region of the complex plane (see e.g., [Kai]) . To be specific we focus on two important classes of linear systems:

domain	transfer function	closed region
discrete	$H(z) = \dfrac{A(z)}{B(z)}$	$\overline{U}^1 = \{z \,:\, \lvert z\rvert \leq 1\}$
continuous	$F(s) = \dfrac{P(s)}{Q(s)}$	$\overline{A}^1 = \{s \,:\, \text{Re}\, s \geq 0\}$

A discrete time system is BIBO stable if and only if

$$B(z) \neq 0 \text{ for each } z \in \overline{U}^1. \tag{5.1.1}$$

The corresponding condition for BIBO stability of a continuous system is given next

$$Q(s) \neq 0 \text{ for each } s \in \overline{A}^1. \tag{5.1.2}$$

Dealing with 2-D systems one considers:

domain	transfer function	closed region
discrete	$H(z_1, z_2) = \dfrac{A(z_1, z_2)}{B(z_1, z_2)}$	$\overline{U}^2 = \{(z_1, z_2) \; : \; \|z_1\| \leq 1, \; \|z_2\| \leq 1\}$
continuous	$F(s_1, s_2) = \dfrac{P(s_1, s_2)}{Q(s_1, s_2)}$	$\overline{A}^2 = \{(s_1, s_2) \; : \; \text{Re } s_1 \geq 0, \; \text{Re } s_2 \geq 0\}$

In contrast to the one dimensional case *two-dimensional* linear shift-invariant digital filters may be stable even when some of the poles of the corresponding transfer function are located on the boundary of $\overline{U}^2$. For example

$$H(z_1, z_2) = \frac{(1 - z_1)^8 (1 - z_2)^8}{2 - z_1 - z_2}$$

is a transfer function of a BIBO stable digital filter. This surprising result and the example are due to Goodman [Go1]. A continuous counterpart of the result and a description of a special class of BIBO stable analog functions are given by Reddy and Jury [RJ]. For example, it is shown in [RJ], that

$$F(s_1, s_2) = \frac{s_1^2 s_2}{(1 + s_1)\left(s_1 s_2 + \frac{1}{2} s_1 + \frac{1}{2} s_2\right)}$$

is a BIBO stable analog function.

As of today there exist no simple necessary and sufficient BIBO stability conditions for 2–D systems. In order to verify stability of a two dimensional system one has to investigate the impulse response of the system. The investigation may result in complicated and long mathematical analysis (see e.g., [Go1] and [JB]). Furthermore, it is known that the Double Bilinear Transformation (DBT) may transform a *stable* 2–D analog function to an *unstable* 2–D digital transfer function (see e.g., [Go2]). A first example of an *unstable* 2–D analog function transformed by DBT to a *stable* 2–D digital transfer function is provided in [JB].

For a comprehensive survey of multidimensional stability we refer the reader to [J1]. Many differences between general problems of one and multidimensional systems are discussed in detail in the survey paper by Bose [Bos1].

From a practical point of view the conventional definition of the n–D BIBO stability is unnecessarily restrictive in most practical situations. A more relevant and less restrictive definition of the *practical*–BIBO stability is introduced by Agathoklis and Bruton [AB]. The relationship between practical–BIBO stability and the singularities of the corresponding transfer function is also described in [AB]. A technique for practical–BIBO stabilization of n–D linear shift-invariant filters has been reported by Swamy, Roytman and Plotkin [SRP].

5.2. Stability of a Bivariate Polynomial

In this section we consider a bivariate polynomial

$$\mathrm{b}(s_1, s_2) = \tau_{0,0} + \tau_{0,1}s_2 + \tau_{0,2}s_2^2 + \ldots + \tau_{1,0}s_1 + \tau_{1,1}s_1 s_2 + \ldots + \tau_{n_1,n_2}s_1^{n_1}s_2^{n_2}. \quad (5.2.1)$$

Let Ω_1 and Ω_2 be two open domains in the complex plane.

Definition 5.2.1 *The bivariate polynomial* $\mathrm{b}(s_1, s_2)$ *is stable if there exists no root of* $\mathrm{b}(s_1, s_2)$ *located in* $\Omega_1^c \times \Omega_2^c$*, otherwise we say that* $\mathrm{b}(s_1, s_2)$ *is unstable.*

Consider parameterizations $\delta_i : I_{\Omega_i} \to \partial\Omega_i$, $i = 1, 2$ of the boundaries of the domains. We first show that stability of a single bivariate polynomial $\mathrm{b}(s_1, s_2)$ can be verified by checking the following two conditions for each $w_i \in I_{\Omega_i}$, $i = 1, 2$:

- stability condition 1: $\mathrm{b}(s_1, \delta_2(w_2)) \neq 0$ when $s_1 \in \Omega_1^c$,
- stability condition 2: $\mathrm{b}(\delta_1(w_1), s_2) \neq 0$ when $s_2 \in \Omega_2^c$,

Conditions of this type are well known in the literature, see e.g., [Bos2], pp. 168–170, where stability tests are developed for the important special case $\Omega_1^c = \Omega_2^c =$ the unit disk (a continuous counterpart of this discrete result is given on p. 300). The next statement is included for the sake of completeness. The idea of the proof is borrowed from [Bas].

Lemma 5.2.1 *The bivariate polynomial* $\mathrm{b}(s_1, s_2)$ *is stable if and only if the stability conditions hold.*

Proof: If the polynomial $\mathrm{b}(s_1, s_2)$ is stable, then the stability conditions follow immediately from Definition 5.2.1. We next assume that the stability conditions hold, and show that the polynomial is stable. Assume that the degree of $\mathrm{b}(s_1, s_2)$ with respect to s_i is n_i, $i = 1, 2$. Suppose that the bivariate polynomial is unstable, i.e., there exist $s_1^0 \in \Omega_1^c$, and $s_2^0 \in \Omega_2^c$ so that $\mathrm{b}(s_1^0, s_2^0) = 0$. The stability conditions imply that $s_1^0 \in \operatorname{int} \Omega_1^c$ and $s_2^0 \in \operatorname{int} \Omega_2^c$. We write the bivariate polynomial $\mathrm{b}(s_1, s_2)$ in the form

$$\left[\sum_{k_2} \tau_{0,k_2} s_2^{k_2}\right] + \left[\sum_{k_2} \tau_{1,k_2} s_2^{k_2}\right] s_1 + \ldots + \left[\sum_{k_2} \tau_{n_1,k_2} s_2^{k_2}\right] s_1^{n_1}.$$

Let $\phi : [0,1] \to \Omega_2^c$ be a continuous mapping such that $\phi(0) = s_2^0$, $\phi(1) \in \partial\Omega_2^c$, and

$$\sum_{k_2} \tau_{n_1,k_2}[\phi(t)]^{k_2} \neq 0 \text{ for each } t \in (0,1].$$

The last condition simply means that for each t, $0 < t \leq 1$ the degree of the univariate polynomial $\mathbf{p}_t(s) = \mathbf{b}(s, \phi(t))$ is n_1. Since the degree of $\mathbf{p}_0(s)$ is beyond our control we have to take care of the two cases:

Case 1. $\sum_{k_2} \tau_{n_1,k_2}[s_2^0]^{k_2} = \sum_{k_2} \tau_{n_1,k_2}[\phi(0)]^{k_2} \neq 0.$

Case 2. $\sum_{k_2} \tau_{n_1,k_2}[s_2^0]^{k_2} = \sum_{k_2} \tau_{n_1,k_2}[\phi(0)]^{k_2} = 0.$

In the first case for each $t \in [0,1]$ one has exactly n_1 complex roots $s_1(t)$, $s_2(t)$, ..., $s_{n_1}(t)$ of the polynomial $\mathbf{p}_t(s)$. When $t = 0$ one of the roots (say $s_1(0)$) lies in int Ω_1^c. Due to continuous dependence of the roots on the coefficients of the polynomials $t \to s_1(t)$ is a continuous mapping. Stability Condition 2 implies that $s_1(t) \in$ int Ω_1^c for each $t \in [0,1]$. Hence $s_1(1) \in$ int Ω_1^c while $s_2(1) \in \partial\Omega_2^c$. This contradicts stability condition 1, and completes the proof of Case 1.

In the second case, due to Theorem of Hurwitz (see [Ti], p. 119), there exists $\lambda \in (0,1]$ so that at least one root of the polynomial $\mathbf{p}_\lambda(s)$ belongs to int Ω_1^c. The rest of the proof is the same as that of the first case.

In the next section we consider a family of bivariate polynomials centered at a stable nominal bivariate polynomial.

5.3. Stability Radius

In this section we consider a family of bivariate polynomials

$$\mathbf{b}(s_1, s_2, \mathbf{d}) = \sum_{i_1=0, i_2=0}^{i_1=n_1, i_2=n_2} \tau_{i_1,i_2}(\mathbf{d}) s^{i_1} s^{i_2}, \tag{5.3.1}$$

where the polynomial coefficients are affine functions of the perturbation vector $\mathbf{d} \in \mathbf{C}^m$,

$$(\tau_{0,0}(\mathbf{d}), \tau_{0,1}(\mathbf{d}), \ldots, \tau_{n_1,n_2}(\mathbf{d})) = \mathbf{d}\mathcal{T} + \boldsymbol{\tau}^0,$$

$\mathcal{T}$ is an $m \times [(n_1+1) \times (n_2+1)]$ matrix, and $\boldsymbol{\tau}^0 \in \mathbf{C}^{[n_1+1] \times [n_2+1]}$. When $\mathbf{d} = \mathbf{0}$ we obtain the nominal bivariate polynomial

$$\mathbf{b}(s_1, s_2, \mathbf{0}) = \tau_{0,0}^0 + \tau_{0,1}^0 s_2 + \tau_{0,2}^0 s_2^2 + \ldots + \tau_{1,0}^0 s_1 + \tau_{1,1}^0 s_1 s_2 + \ldots + \tau_{n_1,n_2}^0 s_1^{n_1} s_2^{n_2}.$$

The measure of the perturbation $\mathbf{d}$ is given by the weighted l_p norm

$$f(\mathbf{d}) = f(a_0 + \jmath b_0, a_1 + \jmath b_1, \ldots, a_{m-1} + \jmath b_{m-1}) = \left[\sum_{k=0}^{m-1} \left|\frac{a_k}{\alpha_k}\right|^p + \sum_{k=0}^{m-1} \left|\frac{b_k}{\beta_k}\right|^p\right]^{1/p}$$

Definition 5.3.1 *Given two open domains Ω_1, and Ω_2 in the complex plane, a matrix T, and the convex function f, the stability radius of a stable nominal bivariate polynomial* $\mathbf{b}(s_1, s_2, \mathbf{0})$ *is defined by*

$$r_{\mathbf{b}} = \inf\{f(\mathbf{d}) \ : \ \mathbf{d} \in \mathbf{C}^m, \ \textit{and } \mathbf{b}(s_1, s_2, \mathbf{d}) \textit{ is unstable}\}.$$

We next show that the evaluation of the stability radius r can be reduced to evaluation of stability radii for families of univariate polynomials.

5.4. Reduction to the Polynomial Problem

Let w_2 be a fixed element of I_{Ω_2}. The bivariate polynomial

$$\mathbf{b}(s_1, \delta_2(w_2), \mathbf{d}) = \tau_{0,0}(\mathbf{d}) + \tau_{0,1}(\mathbf{d})\delta_2(w_2) + \ldots + \tau_{n_1,n_2}(\mathbf{d}) s_1^{n_1} \delta_2(w_2)^{n_2}$$

can be written as a univariate polynomial in s_1:

$$\begin{aligned} \mathbf{b}(s_1, \delta_2(w_2), \mathbf{d}) \ = \ & \left[\sum_{k_2} \tau_{0,k_2}(\mathbf{d})\delta_2(w_2)^{k_2}\right] + \\ & \left[\sum_{k_2} \tau_{1,k_2}(\mathbf{d})\delta_2(w_2)^{k_2}\right] s_1 + \\ & \ldots + \\ & \left[\sum_{k_2} \tau_{n_1,k_2}(\mathbf{d})\delta_2(w_2)^{k_2}\right] s_1^{n_1}. \end{aligned}$$

For each $s \in \mathbf{C}$ we define the $[(n_1+1)\times(n_2+1)]\times[n_1+1]$ matrix $T_2(s)$ as follows:

$$T_2(s) = T \begin{bmatrix} 1 & 0 & \ldots & 0 \\ s & 0 & \ldots & 0 \\ \ldots & \ldots & \ldots & \ldots \\ s^{n_2} & 0 & \ldots & 0 \\ 0 & 1 & \ldots & 0 \\ 0 & s & \ldots & 0 \\ \ldots & \ldots & \ldots & \ldots \\ 0 & s^{n_2} & \ldots & 0 \\ \ldots & \ldots & \ldots & \ldots \\ 0 & 0 & \ldots & 1 \\ 0 & 0 & \ldots & s \\ \ldots & \ldots & \ldots & \ldots \\ 0 & 0 & \ldots & s^{n_2} \end{bmatrix} \tag{5.4.1}$$

Let

$$\mathbf{p}_2(s,\mathbf{d}) = t_0(\mathbf{d}) + t_1(\mathbf{d})s + \ldots + t_{n_1}(\mathbf{d})s^{n_1},$$

where

$$(t_0(\mathbf{d}), t_1(\mathbf{d}), \ldots, t_{n_1}(\mathbf{d})) = \mathbf{d}T_2(\delta_2(w_2)) + \mathbf{t}^0(\delta_2(w_2)),$$

and

$$\begin{aligned}\mathbf{t}^0(\delta_2(w_2)) &= \left(t_0^0(\delta_2(w_2)), t_1^0(\delta_2(w_2)), t_2^0(\delta_2(w_2)), \ldots, t_{n_1}^0(\delta_2(w_2))\right) \\ &= \left(\left[\sum_{k_2} \tau_{0,k_2}^0 \delta_2(w_2)^{k_2}\right], \left[\sum_{k_2} \tau_{1,k_2}^0 \delta_2(w_2)^{k_2}\right], \ldots, \left[\sum_{k_2} \tau_{n_1,k_2}^0 \delta_2(w_2)^{k_2}\right]\right).\end{aligned}$$

Then one has

$$\mathbf{b}(s_1, \delta_2(w_2), \mathbf{d}) = \mathbf{p}_2(s_1, \mathbf{d}).$$

So, if the zero set of $\mathbf{b}(s_1, \delta_2(w_2), \mathbf{d})$ has to be excluded from $\Omega_1^c \times \Omega_2^c$, then the zero set of $\mathbf{p}_2(s_1, \mathbf{d})$ has to be excluded from Ω_1^c. Hence, a polynomial robust stability problem of the following generic form has to be solved:

Given an open domain Ω, a function f, a matrix T, and a polynomial family

$$\mathbf{p}(s,\mathbf{d}) = t_0(\mathbf{d}) + t_1(\mathbf{d})s + \ldots + t_n(\mathbf{d})s^n, \ \mathbf{t}(\mathbf{d}) = \mathbf{d}T + \mathbf{t}^0 \qquad (5.4.2)$$

centered at a stable nominal polynomial

$$\mathbf{p}(s, \mathbf{0}) = t_0^0 + t_1^0 s + \ldots + t_n^0 s^n$$

find the stability radius r of the nominal polynomial $\mathbf{p}(s,\mathbf{0})$, i.e.

$$r = \inf\{f(\mathbf{d}) \ : \ \mathbf{d} \in \mathbf{C}^m, \text{ and } \mathbf{p}(s,\mathbf{d}) \text{ is unstable}\}. \qquad (5.4.3)$$

When $t_n^0 \neq 0$ the solution for this problem is given in Section 3.4. We next show by an example, that when the degree of the nominal polynomial $\mathbf{p}(s,\mathbf{0})$ is less than k, where $\mathbf{T}^k$ is the last nonzero column of the matrix T, the "zero exclusion" criterion fails (see also [CM]).

Example 5.4.1 Let $\Omega = \{s : |s| < 1\}$, $T = [0,0,1]$, $\mathbf{p}(s,\mathbf{0}) = s$, $\mathbf{d} \in \mathbf{C}$, and $f(\mathbf{d}) = |\mathbf{d}|$. The nominal polynomial $\mathbf{p}(s,\mathbf{0}) = s$ is stable. A perturbed polynomial $\mathbf{p}(s,\mathbf{d}) = s + \mathbf{d}s^2 = s(1+\mathbf{d}s)$. When $0 < |\mathbf{d}| \leq 1$ the perturbed polynomial $\mathbf{p}(s,\mathbf{d})$ is unstable, i.e., the stable nominal polynomial $\mathbf{p}(s,\mathbf{0})$ is "surrounded" by unstable polynomials, and the stability radius is 0. On the other hand, when $\mathbf{p}(e^{\mathbf{j}w},\mathbf{d}) = 0$, one has $|\mathbf{d}| = 1$; that is

$$\inf_{0 \leq w < 2\pi} \inf_{\mathbf{d}} \left\{ f(\mathbf{d}) \ : \ \mathbf{p}(e^{\mathbf{j}w}, \mathbf{d}) = 0 \right\} = 1.$$

This shows that a straightforward application of the "zero exclusion" criterion fails. We next show that when $\deg \mathbf{p}(s,\mathbf{0}) < k$ and Ω^c is an unbounded subset of the complex plane, the nominal polynomial is always surrounded by unstable polynomials, and its stability radius is 0.

Lemma 5.4.1 *Let n_0 be the degree of the nominal stable polynomial $\mathbf{p}(s,\mathbf{0})$, and $\mathbf{T}^k$ is the last nonzero column of the matrix T, i.e.,*

$$\mathbf{t}^0 = (t_0^0, t_1^0, \ldots, t_{n_0}^0, 0, \ldots, 0), \text{ and } T = \left[\mathbf{T}^0, \mathbf{T}^1, \ldots, \mathbf{T}^k, 0, \ldots, \mathbf{0}\right].$$

If Ω^c is an unbounded subset of the complex plane, and $n_0 < k$, then for each positive integer l there exist $s_l \in \Omega^c$, and $\mathbf{d}^l \in \mathbf{C}^m$ such that

1. $f(\mathbf{d}^l) < \frac{1}{l}$.
2. $\mathbf{p}(s_l, \mathbf{d}^l) = 0$.

Proof: The proof is identical to the proof of Lemma 3.1.3.

On the other hand if Ω^c is a bounded set the "zero exclusion criterion" holds.

Lemma 5.4.2 *If Ω^c is a bounded subset of the complex plane whose boundary is parameterized by $\delta : I_\Omega \to \partial\Omega^c$, then*

$$r = \inf\{f(\mathbf{d}) \;:\; \mathbf{d} \in \mathbf{C}^m, \text{ and } \mathbf{p}(s,\mathbf{d}) \text{ is unstable}\} = \inf_w \inf_{\mathbf{d}} \{f(\mathbf{d}) \;:\; \mathbf{p}(\delta(w), \mathbf{d}) = 0\}.$$

Proof: The stability of $\mathbf{p}(s,\mathbf{0})$, and the boundedness of Ω^c imply the existence of a positive bound b so that for each $s \in \Omega^c$ one has $b < |\mathbf{p}(s,\mathbf{0})|$. Since

$$|\mathbf{p}(s,\mathbf{d}) - \mathbf{p}(s,\mathbf{0})| \le \sum_{i=0}^{n} \left|\langle \mathbf{d}, \overline{\mathbf{T}^i}\rangle\right| \cdot \left|s^i\right|$$

there exists a stable neighborhood of the origin in the parameter space, i.e., there exists $\epsilon^* > 0$ such that

$$\forall \mathbf{d} \text{ with } f(\mathbf{d}) < \epsilon^*, \text{ and } \forall s \in \Omega^c \text{ one has } \mathbf{p}(s,\mathbf{d}) \neq 0.$$

Let $\mathbf{p}(s,\mathbf{d})$ be an unstable polynomial. We next show that for each positive ϵ one can find $w \in I_\Omega$ and a perturbation $\mathbf{d}'$ so that

$$\mathbf{p}(\delta(w), \mathbf{d}') = 0, \text{ and } f(\mathbf{d}') \le f(\mathbf{d}) + \epsilon.$$

Consider the polynomial segment $\{\mathbf{p}(s, \lambda\mathbf{d})\}$, $0 \le \lambda \le 1$.

Case 1. $\deg \mathbf{p}(s,\mathbf{0}) \ge \deg \mathbf{p}(s,\mathbf{d})$.

The proof of the case follows from Theorem 3.1.3.

Case 2. $\deg \mathbf{p}(s,\mathbf{0}) < \deg \mathbf{p}(s,\mathbf{d})$.

There exists $\lambda_0 \in (0, \epsilon^*)$ so that

1. the corresponding polynomial $\mathbf{p}(s, \lambda_0 \mathbf{d})$ is stable,

2. deg $\mathbf{p}(s, \lambda_0 \mathbf{d}) \geq$deg $\mathbf{p}(s, \mathbf{d})$.

The rest of the proof is the same as that of Case 1.

5.5. Evaluation of Stability Radius

To verify stability condition 1 (see p. 102) one has to find:

1. whether there exists a point $w_2 \in I_{\Omega_2}$ such that $n < k$, where n is the degree of the polynomial $\mathbf{p}_2(s, \mathbf{0})$, and $\mathbf{T}_2^k$ is the last nonzero column of the matrix $T_2(\delta_2(w_2))$.

2. $\inf\limits_{(w_1,w_2)\in\Omega_1\times\Omega_2} \inf\limits_{\mathbf{d}} \{f(\mathbf{d}) \; : \; \mathbf{b}(\delta_1(w_1), \delta_2(w_2), \mathbf{d}) = 0\}.$

In order to deal with stability condition 2 (see p. 102) we have to introduce additional notations. Let

$$\begin{aligned} \mathbf{p}_1(s, \mathbf{d}) \;\; = \;\; & \left[\sum_{k_1} \tau_{k_1,0}(\mathbf{d})\delta_1(w_1)^{k_1}\right] + \\ & \left[\sum_{k_1} \tau_{k_1,1}(\mathbf{d})\delta_1(w_1)^{k_1}\right] s + \\ & \ldots + \\ & \left[\sum_{k_1} \tau_{k_1,n_2}(\mathbf{d})\delta_1(w_1)^{k_1}\right] s^{n_2}. \end{aligned}$$

For each $s \in \mathbf{C}$ we define the $[(n_1+1) \times (n_2+1)] \times [n_2+1]$ matrix $T_1(s)$ as follows:

$$T_1(s) = \mathcal{T} \begin{bmatrix} 1 & 0 & \ldots & 0 \\ 0 & 1 & \ldots & 0 \\ \ldots & \ldots & \ldots & \ldots \\ s & 0 & \ldots & 1 \\ 0 & s & \ldots & 0 \\ \ldots & \ldots & \ldots & \ldots \\ s^2 & 0 & \ldots & s \\ 0 & s^2 & \ldots & 0 \\ \ldots & \ldots & \ldots & \ldots \\ s^{n_1} & 0 & \ldots & s^2 \\ 0 & s^{n_1} & \ldots & 0 \\ \ldots & \ldots & \ldots & \ldots \\ 0 & 0 & \ldots & s^{n_1} \end{bmatrix} \tag{5.5.1}$$

To verify stability condition 2 one has to find:

1. whether there exists a point $w_1 \in I_{\Omega_1}$ such that $n < k$, where n is the degree of the polynomial $\mathbf{p}_1(s, \mathbf{0})$, and $\mathbf{T}_1^k$ is the last nonzero column of the matrix $T_1(\delta_1(w_1))$.

2. $\inf_{(w_1,w_2)\in\Omega_1\times\Omega_2} \inf_{\mathbf{d}}\{f(\mathbf{d}) \,:\, \mathbf{b}(\delta_1(w_1), \delta_2(w_2), \mathbf{d}) = 0\}$.

Definition 5.5.1 *A point $s_0 \in \mathbf{C}$ is a stability zero point if*

1. $s_0 \in \Omega_i^c$, $i \in \{1,2\}$;

2. $n < k$, where n is the degree of the polynomial $\mathbf{p}_i(s, \mathbf{0})$, and $\mathbf{T}_i^k$ is the last nonzero column of the matrix $T_i(s_0)$.

The formula for the stability radius is given next.

Theorem 5.5.1 Let Ω_1 and Ω_2 be open domains in the complex plane whose boundaries are parameterized by $\delta_i \,:\, I_{\Omega_i} \to \partial\Omega_i$, $i = 1, 2$. For a given $m \times [(n_1+1) \times (n_2+1)]$ matrix $\mathcal{T}$, and a stable nominal bivariate polynomial

$$\mathbf{b}(s_1, s_2, \mathbf{0}) = \tau_{0,0}^0 + \tau_{0,1}^0 s_2 + \tau_{0,2}^0 s_2^2 + \ldots + \tau_{1,0}^0 s_1 + \tau_{1,1}^0 s_1 s_2 + \ldots + \tau_{n_1,n_2}^0 s_1^{n_1} s_2^{n_2}$$

the stability radius is given by

$$r_{\mathbf{b}} = \begin{cases} 0 \text{ if there exist an unbounded } \Omega_i^c \text{ that contains stability zero points,} \\ \inf_{(w_1,w_2)\in\Omega_1\times\Omega_2} \inf_{\mathbf{d}}\{f(\mathbf{d}) \,:\, \mathbf{b}(\delta_1(w_1), \delta_2(w_2), \mathbf{d}) = 0\} \text{ otherwise.} \end{cases}$$

The set of stability zero points is a subset of the set of the roots of the two univariate polynomials:

$$\sum_k \tau_{k,n_2}^0 s^k, \text{ and } \sum_k \tau_{n_1,k}^0 s^k. \tag{5.5.2}$$

In many cases (for example when $\mathcal{T}$ is the identity matrix) the last columns of the matrices $T_1(s)$ and $T_2(s)$ are always nonzero vectors. Then the stability zero points are exactly the roots of the polynomials (5.5.2), and when either Ω_1^c is unbounded and contains at least one root of $\sum \tau_{k,n_2}^0 s^k$, or Ω_2^c is unbounded and contains at least one root of $\sum \tau_{n_1,k}^0 s^k$ the stability radius is 0. Otherwise the evaluation of

$$\inf_{(w_1,w_2)\in\Omega_1\times\Omega_2} r_{\mathbf{b}}(w_1, w_2) = \inf_{(w_1,w_2)\in\Omega_1\times\Omega_2} \inf_{\mathbf{d}} \{f(\mathbf{d}) \,:\, \mathbf{b}(\delta_1(w_1), \delta_2(w_2), \mathbf{d}) = 0\} \tag{5.5.3}$$

represents the main computational burden for evaluation of the stability radius $r_{\mathbf{b}}$. For a fixed pair $w_1 \in I_{\Omega_1}$, and $w_2 \in I_{\Omega_2}$ the condition $\mathbf{b}(\delta_1(w_1), \delta_2(w_2), \mathbf{d}) = 0$ can be written as

$$\mathbf{b}(\delta_1(w_1), \delta_2(w_2), \mathbf{d}) - \mathbf{b}(\delta_1(w_1), \delta_2(w_2), \mathbf{0}) = -\mathbf{b}(\delta_1(w_1), \delta_2(w_2), \mathbf{0}). \tag{5.5.4}$$

Let

$$\boldsymbol{\delta}(w_1, w_2) =$$
$$(1, \delta_2(w_2), \ldots, \delta_2(w_2)^{n_2}, \delta_1(w_1), \delta_1(w_1)\delta_2(w_2), \ldots, \delta_1(w_1)\delta_2(w_2)^{n_2}, \ldots, \delta_1(w_1)^{n_1}\delta_2(w_2)^{n_2})^t.$$

Equation (5.5.4) can now be written as

$$\langle \mathbf{d}, \mathbf{s}(w_1, w_2)\rangle = 1, \text{ where } \mathbf{s}(w_1, w_2) = \overline{-\mathcal{T}\left[\frac{\boldsymbol{\delta}(w_1, w_2)}{\mathbf{b}(\delta_1(w_1), \delta_2(w_2), \mathbf{0})}\right]}. \tag{5.5.5}$$

Thus $r_{\mathbf{b}}(w_1, w_2)$ is given by

$$r_{\mathbf{b}}(w_1, w_2) = \inf\{f(\mathbf{d}) : \mathbf{d} \in \mathbf{C}^m, \langle \mathbf{d}, \mathbf{s}(w_1, w_2)\rangle = 1\}. \tag{5.5.6}$$

This optimization problem has been solved in Section 3.4.

5.6. Example: Hurwitz Stability of Interval Bivariate Polynomials

In this section we evaluate the stability radius of a family of bivariate polynomials for the following special case: $\mathcal{T}$ is the identity matrix, $\Omega_1 = \Omega_2$ =the left half plane. The parameters $\mathbf{d}$ are coefficients of the polynomials, and it is convenient to measure the distance between two bivariate polynomials

$$\mathbf{b}^0(s_1, s_2) = \tau_{0,0}^0 + \tau_{0,1}^0 s_2 + \ldots + \tau_{n_1,n_2}^0 s_1^{n_1} s_2^{n_2}$$

and

$$\mathbf{b}(s_1, s_2) = \tau_{0,0} + \tau_{0,1} s_2 + \ldots + \tau_{n_1,n_2} s_1^{n_1} s_2^{n_2}$$

by

$$\left[\sum_{k_1,k_2}\left|\frac{a_{k_1,k_2} - a_{k_1,k_2}^0}{\alpha_{k_1,k_2}}\right|^p + \sum_{k_1,k_2}\left|\frac{b_{k_1,k_2} - b_{k_1,k_2}^0}{\beta_{k_1,k_2}}\right|^p\right]^{1/p},$$

where $a_{k_1,k_2} + \boldsymbol{j} b_{k_1,k_2} = \tau_{k_1,k_2}$. In this example we choose $p = \infty$. First we have to verify that the polynomials

$$\sum_k \tau_{n_1,k}^0 s^k, \text{ and } \sum_k \tau_{k,n_2}^0 s^k \tag{5.6.1}$$

are Hurwitz stable, i.e., all their roots are located in the open left half plane (if this is not the case the right half plane contains stability zero points, and the stability radius $r_{\mathbf{b}} = 0$). If the polynomials (5.6.1) are Hurwitz stable, then the stability radius is the greatest lower bound of the function $r(w_1, w_2)$ defined by (5.5.6). In order to evaluate $r_b(w_1, w_2)$ one has to have formulas for coordinates of the vectors **u**, and **v** (see Section 3.4)

$$\begin{aligned} \mathbf{u} &= (\alpha_{0,0}\phi_{0,0}, \ldots, \alpha_{n_1,n_2}\phi_{n_1,n_2}, \beta_{0,0}\psi_{0,0}, \ldots, \beta_{n_1,n_2}\psi_{n_1,n_2}), \\ \mathbf{v} &= (-\alpha_{0,0}\psi_{0,0}, \ldots, -\alpha_{n_1,n_2}\psi_{n_1,n_2}, \beta_{0,0}\phi_{0,0}, \ldots, \beta_{n_1,n_2}\phi_{n_1,n_2}). \end{aligned} \tag{5.6.2}$$

A straightforward substitution into (5.5.5) yields the following expressions for ψ_{k_1,k_2}, and ϕ_{k_1,k_2}:

$$|\mathbf{b}^0(jw_1, jw_2)|^2 \psi_{k_1,k_2} = \begin{cases} -w_1^{k_1} w_2^{k_2} \operatorname{Im} \mathbf{b}^0(jw_1, jw_2), & \text{when } k_1 + k_2 = 4m, \\ w_1^{k_1} w_2^{k_2} \operatorname{Re} \mathbf{b}^0(jw_1, jw_2), & \text{when } k_1 + k_2 = 4m+1, \\ w_1^{k_1} w_2^{k_2} \operatorname{Im} \mathbf{b}^0(jw_1, jw_2), & \text{when } k_1 + k_2 = 4m+2, \\ -w_1^{k_1} w_2^{k_2} \operatorname{Re} \mathbf{b}^0(jw_1, jw_2), & \text{when } k_1 + k_2 = 4m+3, \end{cases}$$

and

$$|\mathbf{b}^0(jw_1, jw_2)|^2 \phi_{k_1,k_2} = \begin{cases} -w_1^{k_1} w_2^{k_2} \operatorname{Re} \mathbf{b}^0(jw_1, jw_2), & \text{when } k_1 + k_2 = 4m, \\ -w_1^{k_1} w_2^{k_2} \operatorname{Im} \mathbf{b}^0(jw_1, jw_2), & \text{when } k_1 + k_2 = 4m+1, \\ w_1^{k_1} w_2^{k_2} \operatorname{Re} \mathbf{b}^0(jw_1, jw_2), & \text{when } k_1 + k_2 = 4m+2, \\ w_1^{k_1} w_2^{k_2} \operatorname{Im} \mathbf{b}^0(jw_1, jw_2), & \text{when } k_1 + k_2 = 4m+3. \end{cases}$$

Let

$$T_\infty(w_1, w_2) = \sum_{k_1+k_2 \text{ is odd}} \alpha_{k_1,k_2} |w_1|^{k_1} |w_2|^{k_2} + \sum_{k_1+k_2 \text{ is even}} \beta_{k_1,k_2} |w_1|^{k_1} |w_2|^{k_2},$$

and

$$S_\infty(w_1, w_2) = \sum_{k_1+k_2 \text{ is even}} \alpha_{k_1,k_2} |w_1|^{k_1} |w_2|^{k_2} + \sum_{k_1+k_2 \text{ is odd}} \beta_{k_1,k_2} |w_1|^{k_1} |w_2|^{k_2}.$$

A straightforward application of (3.4.8) shows that

$$r_b(w_1, w_2) = \max\left\{ \frac{|\operatorname{Re} \mathbf{b}^0(jw_1, jw_2)|}{S_\infty(w_1, w_2)}, \frac{|\operatorname{Im} \mathbf{b}^0(jw_1, jw_2)|}{T_\infty(w_1, w_2)} \right\}.$$

The stability radius is given by:

$$r_b = \inf_{w_1,w_2} r_b(w_1, w_2) = \inf_{w_1,w_2} \left[\max\left\{ \frac{|\operatorname{Re} \mathbf{b}^0(jw_1, jw_2)|}{S_\infty(w_1, w_2)}, \frac{|\operatorname{Im} \mathbf{b}^0(jw_1, jw_2)|}{T_\infty(w_1, w_2)} \right\} \right].$$

5.7. Quasipolynomials with Commensurate Delays and Zero Exclusion Criterion

The remainder of the chapter and the next chapter are concerned with Hurwitz stability of time delay systems. In this section we consider a quasipolynomial family with

K interval commensurate delays

$$\mathbf{Q} = \left\{ \mathbf{q}(s, \mathbf{d}, \tau) = \sum_{n,k=0}^{N,K} t_{n,k}(\mathbf{d}) s^n e^{-k\tau s} \; : \; \mathbf{d} \in \mathbf{D}, \; \tau \in [\underline{\tau}, \overline{\tau}] \right\}, \tag{5.7.1}$$

where $\mathbf{D}$ is a compact and convex subset of $\mathbf{C}^{(N+1)\times(K+1)}$, and $t_{n,k}(\mathbf{d})$ are continuous (complex or real valued) functions of $\mathbf{d}$. The main result of the section is the zero exclusion criterion for the quasipolynomial family (5.7.1).

Theorem 5.7.1 ("zero exclusion" criterion.) Suppose that $\mathbf{q}(s, \mathbf{d}^0, \tau^0)$ is a stable quasipolynomial. If there exists an unstable quasipolynomial $\mathbf{q}(s, \mathbf{d}^1, \tau^1)$, then there exists a quasipolynomial $\mathbf{q}(s, \mathbf{d}, \tau)$, and $w \in \mathbf{R}$ such that $\mathbf{q}(jw, \mathbf{d}, \tau) = 0$.

The zeros of a quasipolynomial in general are not continuous in $(\mathbf{d}, \tau)$ (see e.g., [AH], [ADM], [Si]). Hence, for the proof of the theorem which follows Xin and Feng [XF], we will need a number of auxiliary assumptions and results.

Let $\psi_n(s, \mathbf{d}, \tau) = \sum_{k=0}^{K} t_{n,k}(\mathbf{d}) e^{-k\tau s}$, then

$$\mathbf{q}(s, \mathbf{d}, \tau) = \psi_0(s, \mathbf{d}, \tau) + \psi_1(s, \mathbf{d}, \tau)s + \psi_2(s, \mathbf{d}, \tau)s^2 + \ldots + \psi_N(s, \mathbf{d}, \tau)s^N. \tag{5.7.2}$$

The leading coefficient in (5.7.2) is

$$\psi_N(s, \mathbf{d}, \tau) = t_{N,0}(\mathbf{d}) + t_{N,1}(\mathbf{d})e^{-\tau s} + t_{N,2}(\mathbf{d})e^{-2\tau s} + \ldots + t_{N,K}(\mathbf{d})e^{-K\tau s}. \tag{5.7.3}$$

In order to avoid the "degree dropping" we assume the following.

Assumption 5.7.1 For each $\mathbf{d} \in \mathbf{D}$, $\tau \in [\underline{\tau}, \overline{\tau}]$, and s with $\mathrm{Re}\, s \geq 0$ one has

$$\psi_N(s, \mathbf{d}, \tau) \neq 0, \text{ and } t_{N,0}(\mathbf{d}) \neq 0.$$

There is a variety of different conditions under which the zero exclusion criterion is valid for quasipolynomial families. For example, an elegant modification of Assumption 5.7.1 due to V.L. Kharitonov is given by Assumption 6.7.2, page 133. For additional discussion of conditions that yield the zero exclusion we address the interested reader to [HIT].

When $\mathrm{Re}\, s \geq 0$ one has $|\psi_n(s, \mathbf{d}, \tau)| \leq \sum_{k=0}^{K} \left| t_{n,k}(\mathbf{d}) e^{-k\tau s} \right| \leq \sum_{k=0}^{K} |t_{n,k}(\mathbf{d})|$. Continuity of $t_{n,k}$ along with compactness of $\mathbf{D}$ yields the following.

Observation 5.7.1 There exists $M > 0$ such that

$$M > |\psi_n(s, \mathbf{d}, \tau)| \text{ for each } \mathbf{d} \in \mathbf{D}, \; \tau \in [\underline{\tau}, \overline{\tau}], \; s \text{ with } \mathrm{Re}\, s \geq 0, \text{ and } n = 0, \ldots, N.$$

Let $m_0 = \inf\{|t_{N,0}(\mathbf{d})| \,:\, \mathbf{d} \in \mathbf{D}\}$. According to Assumption 5.7.1 one has $m_0 > 0$. When ρ is a large positive number and Re $s > \rho$ the first term dominates in (5.7.3), that is:

Observation 5.7.2 There exists $\rho > 0$ such that $|\psi_N(s, \mathbf{d}, \tau)| \geq \frac{1}{2} m_0$ when Re $s > 0$.

The next statement deals with the strip $\{s \,:\, 0 \leq \text{Re } s \leq \rho\}$.

Lemma 5.7.1 *For each $\rho > 0$ there exists $m_\rho > 0$ such that*

$$|\psi_N(s, \mathbf{d}, \tau)| \geq m_\rho \quad \textit{for each } \mathbf{d} \in \mathbf{D},\ \tau \in [\underline{\tau}, \overline{\tau}], \quad \textit{and } s \textit{ with } 0 \leq Re\, s \leq \rho.$$

Proof: For each $0 \leq \sigma \leq \rho$, $0 \leq w \leq T = \dfrac{2\pi}{\underline{\tau}}$, $\mathbf{d} \in \mathbf{D}$, and $\tau \in [\underline{\tau}, \overline{\tau}]$ let

$$\phi(\sigma, w, \mathbf{d}, \tau) = t_{N,0}(\mathbf{d}) + t_{N,1}(\mathbf{d})e^{-\tau(\sigma + \boldsymbol{j} w)} + t_{N,2}(\mathbf{d})e^{-2\tau(\sigma+\boldsymbol{j}w)} + \ldots + t_{N,K}(\mathbf{d})e^{-K\tau(\sigma+\boldsymbol{j}w)}.$$

The continuous function $|\phi|$ is defined in a compact set. Let

$$m_\rho = \min|\phi| = |\phi(\sigma^0, w^0, \mathbf{d}^0, \tau^0)|.$$

Since $\phi(\sigma^0, w^0, \mathbf{d}^0, \tau^0) = \psi_N(\sigma^0 + \boldsymbol{j}w^0, \mathbf{d}^0, \tau^0) \neq 0$ one has $m_\rho > 0$. Let $s = \sigma^s + \boldsymbol{j}w^s$ with $0 \leq \sigma^s \leq \rho$. Since

$$\psi_N(\sigma^s + \boldsymbol{j}w^s, \mathbf{d}, \tau) = \psi_N\left(\sigma^s + \boldsymbol{j}\left[w^s + \frac{2\pi}{\tau}m\right], \mathbf{d}, \tau\right), \quad m = 0, \pm 1, \ldots$$

one can pick m such that

$$0 \leq w = w^s + \frac{2\pi}{\tau}m \leq \frac{2\pi}{\tau} \leq \frac{2\pi}{\underline{\tau}},$$

and $|\psi_N(s, \mathbf{d}, \tau)| = |\psi_N(\sigma^s + \boldsymbol{j}w, \mathbf{d}, \tau)| = |\phi(\sigma^s, w, \mathbf{d}, \tau)| \geq m_\rho$.

Corollary 5.7.1 *For a positive ρ promised in Observation 5.7.2 let $m = \min\left\{\dfrac{m_0}{2}, m_\rho\right\}$. Then*

$$|\psi_N(s, \mathbf{d}, \tau)| \geq m \textit{ for each } \mathbf{d} \in \mathbf{D},\ \tau \in [\underline{\tau}, \overline{\tau}],\ s \textit{ with } Re\, s \geq 0.$$

Let $\mathbf{B}(\epsilon)$ be a quasipolynomial box centered at $\mathbf{q}(s, \mathbf{d}^0, \tau^0)$, i.e.,

$$\mathbf{B}(\epsilon) = \left\{\mathbf{q}(s, \mathbf{d}, \tau) \,:\, \mathbf{d} \in \mathbf{D},\ \tau \in [\underline{\tau}, \overline{\tau}],\ \max\{|\mathbf{d} - \mathbf{d}^0|_\infty, |\tau - \tau^0|\} \leq \epsilon\right\}.$$

Lemma 5.7.2 *Suppose that $\mathbf{q}(s, \mathbf{d}^0, \tau^0)$ is a stable quasipolynomial. There exists $\epsilon > 0$ such that the box $\mathbf{B}(\epsilon)$ is stable.*

Proof: Let $\rho > \frac{M}{m}N$. If $|s| \geq \rho$ one has

$$\begin{aligned}|\mathbf{q}(s,\mathbf{d},\tau)| &\geq |\psi_N(s,\mathbf{d},\tau)||s|^N - \left[|\psi_{N-1}(s,\mathbf{d},\tau)|\,|s|^{N-1} + \ldots + |\psi_0(s,\mathbf{d},\tau)|\right] \\ &\geq |s|^N\left[m - \frac{MN}{\rho}\right] > 0.\end{aligned}$$

The existence of the ϵ follows from compactness of the set $\{s \ : \ |s| \leq \rho\}$, and continuity of $\mathbf{q}(s,\mathbf{d},\tau)$ with respect to $(s,\mathbf{d},\tau)$.

We are ready now to complete the proof of Theorem 5.7.1. Consider the quasipolynomial family

$$\mathbf{q}(s,\mathbf{d}^\lambda,\tau^\lambda) \text{ where } \mathbf{d}^\lambda = (1-\lambda)\mathbf{d}^0 + \lambda\mathbf{d}^1, \text{ and } \tau^\lambda = (1-\lambda)\tau^0 + \lambda\tau^1.$$

Let

$$\lambda_0 = \inf\left\{\lambda \ : \ 0 \leq \lambda \leq 1, \ \mathbf{q}(s,\mathbf{d}^\lambda,\tau^\lambda) \text{ is unstable}\right\}. \tag{5.7.4}$$

According to Lemma 5.7.2 the quasipolynomial $\mathbf{q}(s,\mathbf{d}^{\lambda_0},\tau^{\lambda_0})$ is unstable. Due to Assumption 5.7.1 the unstable roots of $\mathbf{q}(s,\mathbf{d}^{\lambda_0},\tau^{\lambda_0})$ are bounded. Hence the set of the unstable roots is finite. Relation (5.7.4) and Theorem of Hurwitz yield the existence of an unstable root s with Re $s = 0$. This completes the proof of the theorem.

5.8. Diamond Quasipolynomials: Eight Edges Theorem

In this section we present an example of a complex coefficient weighted diamond quasipolynomial family with interval delays first considered in [HKZ1] . Due to a special choice of the weights the stability criterion derived in this section has a particularly simple form.

Consider a complex coefficients diamond quasipolynomial family with K commensurate delays

$$\mathbf{Q} = \left\{\mathbf{q}(s,\mathbf{x},\mathbf{y},\tau) = \sum_{n,k=0}^{N,K}\left(\left[\underline{a}_{n,k} + \alpha_{n,k}x_{n,k}\right] + j\left[\underline{b}_{n,k} + \beta_{n,k}y_{n,k}\right]\right)s^n e^{-k\tau s}\right\}, \tag{5.8.1}$$

$$\mathbf{D} = \left\{(\mathbf{x},\mathbf{y}) \ : \ \sum_{n,k}|x_{n,k}| + \sum_{n,k}|y_{n,k}| \leq 1\right\}, \text{ and } 0 < \underline{\tau} \leq \tau \leq \overline{\tau}. \tag{5.8.2}$$

Let

$$\begin{aligned}&\alpha_{0,0} = \alpha_{N,0} = \beta_{0,0} = \beta_{N,0} = r\gamma \quad r \geq \sqrt{2}, \\ &\gamma = \max\{\alpha_{n,k},\ \beta_{n,k}\} \qquad (n,k) \neq (0,0),(N,0).\end{aligned} \tag{5.8.3}$$

Assumption 5.7.1 in this particular case reads:

Assumption 5.8.1 For each $(\mathbf{x},\mathbf{y}) \in \mathbf{D}$, $\tau \in [\underline{\tau},\overline{\tau}]$, and s with $\operatorname{Re} s \geq 0$ one has

$$\sum_{k=0}^{K}\left(\left[\underline{a}_{N,k}+\alpha_{N,k}x_{N,k}\right]+\boldsymbol{j}\left[\underline{b}_{N,k}+\beta_{N,k}y_{N,k}\right]\right)e^{-k\tau s}\neq 0,$$

and

$$\left[\underline{a}_{N,0}+\alpha_{N,0}x_{N,0}\right]+\boldsymbol{j}\left[\underline{b}_{N,0}+\beta_{N,0}y_{N,0}\right]\neq 0.$$

For fixed $w \in \mathbf{R}$ and $\tau \in [\underline{\tau},\overline{\tau}]$ let

$$\mathcal{Q}_{w,\tau}=\{\mathbf{q}(\boldsymbol{j}w,\mathbf{x},\mathbf{y},\tau)\ :\ (\mathbf{x},\mathbf{y})\in\mathbf{D}\}. \tag{5.8.4}$$

It is easy to see that regardless of τ the value set $\mathcal{Q}_{0,\tau}$ is a "rotated" square centered at

$$z_0=\mathbf{q}(0,\mathbf{0},\mathbf{0},0)=\sum_{k=0}^{K}\underline{a}_{0,k}+\boldsymbol{j}\sum_{k=0}^{K}\underline{b}_{0,k}.$$

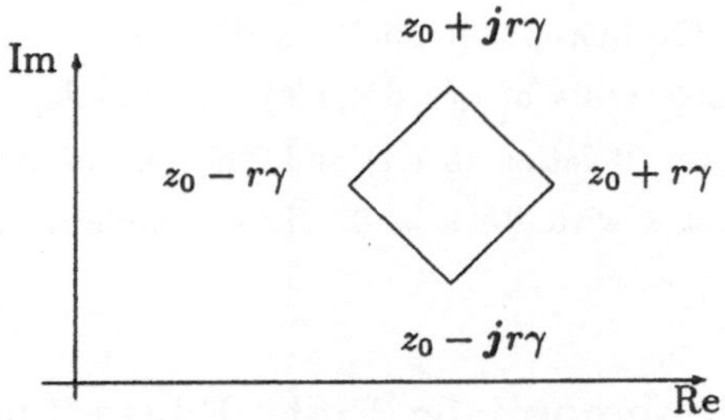

The "rotated" square does not contain the origin if and only if $\left|\sum_{k=0}^{K}\underline{a}_{0,k}\right|+\left|\sum_{k=0}^{K}\underline{b}_{0,k}\right|>r\gamma$. If $0\notin\mathcal{Q}_{0,\tau}$, then the quasipolynomial family is stable if and only if the boundary of $\mathcal{Q}_{w,\tau}$ does not contain the origin for each $w\in\mathbf{R}$ and $\tau\in[\underline{\tau},\overline{\tau}]$. For fixed $w\in\mathbf{R}$ and $\tau\in[\underline{\tau},\overline{\tau}]$ the image of the diamond $\mathbf{D}$ under the mapping

$$(\mathbf{x},\mathbf{y})\rightarrow\mathbf{q}(\boldsymbol{j}w,\mathbf{x},\mathbf{y},\tau) \tag{5.8.5}$$

is a convex polygon in the complex plane. Due to the particular choice of the weights (5.8.3) the polygon is a "rotated" square centered at $z_c=z_c(\boldsymbol{j}w,\tau)=\mathbf{q}(\boldsymbol{j}w,0,0,\tau)$

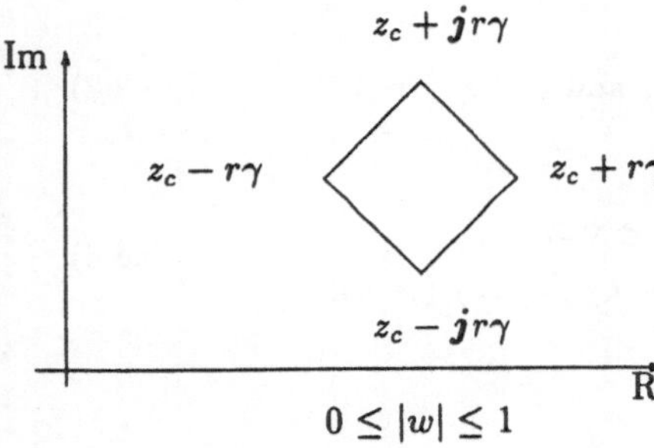

$0\leq|w|\leq 1$

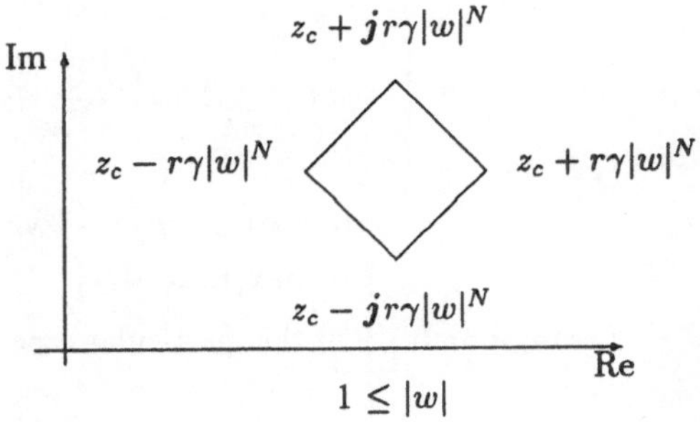

$1\leq|w|$

The boundary of the value set $\mathcal{Q}_{w,\tau}$ is made up of four segments

$$\lambda\,[\mathbf{q}(\boldsymbol{j}w,\mathbf{0},\mathbf{0},\tau)\pm r\gamma]+(1-\lambda)\,[\mathbf{q}(\boldsymbol{j}w,\mathbf{0},\mathbf{0},\tau)\pm \boldsymbol{j}r\gamma]\,,\qquad \text{when } |w|\le 1,$$
$$\lambda\left[\mathbf{q}(\boldsymbol{j}w,\mathbf{0},\mathbf{0},\tau)\pm r\gamma w^N\right]+(1-\lambda)\left[\mathbf{q}(\boldsymbol{j}w,\mathbf{0},\mathbf{0},\tau)\pm \boldsymbol{j}r\gamma w^N\right],\quad \text{when } 1\le |w|.$$

The robust stability criterion for the family (5.8.1) is given next.

Theorem 5.8.1 The family (5.8.1) is stable under the diamond constraint (5.8.2), if and only if

1. $\left|\sum_{k=0}^{K} \underline{a}_{0,k}\right| + \left|\sum_{k=0}^{K} \underline{b}_{0,k}\right| > r\gamma.$
2. The *eight* two parameter quasipolynomial families

$$\lambda\,[\mathbf{q}(s,\mathbf{0},\mathbf{0},\tau)\pm r\gamma] \quad + \quad (1-\lambda)\,[\mathbf{q}(s,\mathbf{0},\mathbf{0},\tau)\pm \boldsymbol{j}r\gamma]\,,$$
$$\lambda\left[\mathbf{q}(s,\mathbf{0},\mathbf{0},\tau)\pm r\gamma s^N\right] \quad + \quad (1-\lambda)\left[\mathbf{q}(s,\mathbf{0},\mathbf{0},\tau)\pm \boldsymbol{j}r\gamma s^N\right],$$

where

$$0\le\lambda\le 1, \text{ and } \underline{\tau}\le\tau\le\overline{\tau}$$

are stable.

The next chapter is devoted to robust stability of uncertain quasipolynomial families with general (not necessarily commensurate) delays. First, we introduce a number of assumptions under which the "zero exclusion" criterion holds. Next, we provide computationally tractable necessary and sufficient robust stability conditions for quasipolynomial families with interval coefficients and interval delays. This result is the main contribution of the chapter. Finally, we present a recent quasipolynomial "extreme point" result due to Kharitonov and Zhabko [KZ]. An application of the result to diamond quasipolynomial families leads to a significant simplification of robust stability conditions given in Theorem 5.8.1.

It is known that there exist time–delay systems with *unbounded* solutions and *Hurwitz stable* characteristic quasipolynomials (see e.g. [Bru]). In many practical applications it is important to guarantee that the real parts of the roots of a characteristic quasipolynomial do not exceed $-\epsilon$ for some positive ϵ. It turns out that this condition follows from Hurwitz stability of an uncertain quasipolynomial family and each one of the assumptions (reported in this book) under which the "zero exclusion" criterion holds. For details we refer the reader to [HKZ2].

Chapter 6

Uncertain Time–Delay Systems

Stability conditions for time–delay systems are of great importance for industrial applications. Delays often occur in the transmission of information or material between different parts of a system. Transportation systems, communication systems, chemical processing systems, metallurgical processing systems, environmental systems and power systems are examples of time–delay systems (see e.g., [MZJ], [St]). The mathematical formulation of a time–delay system results in a system of delay–differential equations.

The stability analysis of a time–delay system is based on investigation of the root location region for the characteristic quasipolynomial. A fundamental result concerning stability of a quasipolynomial is due to L.S. Pontryagin [P]. A significant research effort has been devoted to robust stability criterion for quasipolynomial families. Fu, Olbrot and Polis [FOP1] , [FOP2] generalized the celebrated Edge Theorem to quasipolynomial families with *constant* delays and coefficients depending affinely on parameters. Barmish and Shi [BS2] investigated robust stability of quasipolynomial families with coefficients depending affinely on parameters and *interval* delays. An application of a frequency domain technique reduces the original robust stability problem to a global minimization problem in a finite dimensional Euclidean space. The latter problem, in general, is very difficult to solve. An efficient algorithm for handling the minimization problem remains to be an open problem. Tsypkin and Fu [TF] proposed a graphical test for quasipolynomial families with *one* interval delay. Boese [Boe1] derived necessary and sufficient stability conditions for quasipolynomials with interval coefficients and *one* interval delay. Computationally tractable technique that handles quasipolynomial families with *constant* coefficients and *interval* delays is reported in [Boe2].

In this chapter following Kogan and Leizarowitz [KL1] we derive computationally tractable necessary and sufficient robust stability conditions for interval quasipolynomial families.

6.1. Time–Delay Systems and Zero Exclusion Criterion

In this section we consider a quasipolynomial family with K independent interval delays

$$\mathbf{Q} = \left\{ \mathbf{q}(s, \mathbf{d}, \boldsymbol{\tau}) = \sum_{n,k=0}^{N,K} t_{n,k}(\mathbf{d}) s^n e^{-\tau_k s} \ : \ \mathbf{d} \in \mathbf{D}, \ \begin{array}{c} \tau_k \in [\underline{\tau}_k, \overline{\tau}_k] \\ k = 1, \ldots, K \end{array} \right\}, \tag{6.1.1}$$

where $\mathbf{D}$ is a compact convex set in $\mathbf{C}^m$, and $t_{n,k} \ : \ \mathbf{C}^m \to \mathbf{C}$ are continuous mappings. We first show by an example that when delays are incommensurable the distance from a stable quasipolynomial to unstable quasipolynomials may be zero even when Assumption 5.7.1 holds.

Example 6.1.1 Consider the quasipolynomial family with two delays that contains a single quasipolynomial

$$\mathbf{Q} = \left\{ \mathbf{q}(s, \mathbf{d}, \boldsymbol{\tau}) = 2 + e^{-s} + e^{-\sqrt{2}s}, \ \mathbf{D} = \{0\}, \ \tau_1 = 1, \ \tau_2 = \sqrt{2} \right\}.$$

In the sequel we denote the quasipolynomial by $\mathbf{q}(s)$. Note that while $\inf_w |\mathbf{q}(\boldsymbol{j}w)| = 0$ the polynomial $\mathbf{q}(s)$ is stable. Indeed, since $|e^{-\boldsymbol{j}w}| = |e^{-\boldsymbol{j}\sqrt{2}w}| = 1$, the condition $\mathbf{q}(\boldsymbol{j}w) = 0$ would imply $e^{-\boldsymbol{j}w} = e^{-\boldsymbol{j}\sqrt{2}w} = -1$, and would lead to

$$w \neq 0, \ -w = \pi + 2\pi k, \ -\sqrt{2}w = \pi + 2\pi n, \text{ and } \sqrt{2} = \frac{2n+1}{2k+1}. \tag{6.1.2}$$

This contradiction shows that $\mathbf{q}(\boldsymbol{j}w)$ does not vanish. For $s = \sigma + \boldsymbol{j}w$ with $\sigma > 0$ one has

$$|\mathbf{q}(s)| \geq 2 - e^{-\sigma} - e^{-\sigma\sqrt{2}} > 0.$$

This shows that $\mathbf{q}(s)$ is stable. On the other hand let $\epsilon > 0$ be given. Pick $\overline{\alpha} > 0$ such that

$$\cos(\pi - \overline{\alpha}) < -1 + \frac{\epsilon}{2}.$$

There exist integers k and n such that

$$0 < \frac{2\pi}{1+\sqrt{2}} \left[\frac{1}{\sqrt{2}} - \frac{1}{2} + \sqrt{2}k - n \right] = \alpha < \overline{\alpha}.$$

If $w_\epsilon = \pi + 2\pi k - \alpha$, then $\sqrt{2}w_\epsilon = \pi + 2\pi n + \alpha$, $\mathbf{q}(\boldsymbol{j}w_\epsilon)$ is real, and

$$0 < \mathbf{q}(\boldsymbol{j}w_\epsilon) = 2 + e^{-\boldsymbol{j}w_\epsilon} + e^{-\boldsymbol{j}\sqrt{2}w_\epsilon} < \epsilon.$$

Furthermore, the quasipolynomial $\mathbf{q}_\delta(s) = [2 - \delta] + e^{-s} + e^{-\sqrt{2}s}$ with $\delta = 2 - \mathbf{q}(\boldsymbol{j}w_\epsilon)$ vanishes at $\boldsymbol{j}w_\epsilon$.

The example shows that assumptions stronger than Assumption 5.7.1 are needed if one wants to generalize Lemma 5.7.1 to time delay systems with incommensurable delays . We write the quasipolynomial $\mathbf{q}(s, \mathbf{d}, \boldsymbol{\tau})$ as

$$\psi_0(s, \mathbf{d}) + \psi_1(s, \mathbf{d})e^{-\tau_1 s} + \psi_2(s, \mathbf{d})e^{-\tau_2 s} + \ldots + \psi_K(s, \mathbf{d})e^{-\tau_K s} \tag{6.1.3}$$

where the polynomials $\psi_k(s, \mathbf{d})$ are given by $\psi_k(s, \mathbf{d}) = \sum_{n=0}^{N} t_{n,k}(\mathbf{d})s^n$. According to Pontryagin's necessary stability condition [P] a stable quasipolynomial has a principal term. The following adjustment of Assumption 5.7.1 is motivated by this condition.

Assumption 6.1.1 At least one quasipolynomial $\mathbf{q}(s, \mathbf{d}^0, \boldsymbol{\tau}^0) \in \mathbf{Q}$ is stable. For each $\mathbf{d} \in \mathbf{D}$, and $k = 1, \ldots, K$ one has

$$\deg \psi_0(s, \mathbf{d}) = d_0 > \deg \psi_k(s, \mathbf{d}).$$

The value set $\mathcal{Q}_w$ which is defined by

$$\mathcal{Q}_w = \{\mathbf{q}(jw, \mathbf{d}, \boldsymbol{\tau}) \ : \ \mathbf{d} \in \mathbf{D}, \ \text{ and } \underline{\tau}_k \leq \tau_k \leq \overline{\tau}_k, \ k = 1, \ldots, K\} \tag{6.1.4}$$

plays an important role in the development presented next.

Theorem 6.1.1 ("zero exclusion" criterion). The quasipolynomial family $\mathbf{Q}$ is stable if and only if

$$0 \notin \mathcal{Q}_w \text{ for each } w \in \mathbf{R}. \tag{6.1.5}$$

In many cases verification of (6.1.5) requires a significant computational effort. If for some w_0 the value set $\mathcal{Q}_{w_0}$ has a simple shape, then requirements of Theorem 6.1.1 can be relaxed as follows.

Theorem 6.1.2 Suppose that $0 \notin \mathcal{Q}_{w_0}$. The quasipolynomial family $\mathbf{Q}$ is stable if and only if

$$0 \notin \partial\mathcal{Q}_w \text{ for each } w \in \mathbf{R}. \tag{6.1.6}$$

6.1.1. Proof of Zero Exclusion Criterion

In this subsection we assume the existence of a stable $\mathbf{q}(s, \mathbf{d}^0, \boldsymbol{\tau}^0)$ and unstable $\mathbf{q}(s, \mathbf{d}^1, \boldsymbol{\tau}^1)$ elements of $\mathbf{Q}$, and show the existence of a quasipolynomial $\mathbf{q}(s, \mathbf{d}, \boldsymbol{\tau})$ and $w \in \mathbf{R}$ such that $\mathbf{q}(jw, \mathbf{d}, \boldsymbol{\tau}) = 0$ (see Theorem 6.1.3).

We denote elements of the quasipolynomial segment

$$\left\{\mathbf{q}\left(s, (1-\lambda)\,\mathbf{d}^0 + \lambda\mathbf{d}^1, (1-\lambda)\,\boldsymbol{\tau}^0 + \lambda\boldsymbol{\tau}^1\right),\ 0 \leq \lambda \leq 1\right\} \text{ by } \mathbf{q}\left(s, \mathbf{d}^\lambda, \boldsymbol{\tau}^\lambda\right). \tag{6.1.7}$$

Lemma 6.1.1 *There exists $\epsilon > 0$ such that the quasipolynomial box*

$$\left\{\mathbf{q}(s,\mathbf{d},\boldsymbol{\tau}) \;:\; \left|\mathbf{d}-\mathbf{d}^0\right|_\infty \le \epsilon,\; \left|\boldsymbol{\tau}-\boldsymbol{\tau}^0\right|_\infty \le \epsilon\right\} \tag{6.1.8}$$

centered at $\mathbf{q}(s,\mathbf{d}^0,\boldsymbol{\tau}^0)$ is stable.

Proof: There exists $\rho > 0$ so that $\psi_0(s,\mathbf{d}^0) \neq 0$ when $|s| > \rho$. Furthermore, due to the degree condition of Assumption 6.1.1 one can choose ρ so large that for each $\mathbf{q} \in \mathbf{Q}$ and s with Re $s \ge 0$, and $|s| > \rho$ the following holds

$$\left|\psi_0(s,\mathbf{d}^0)\right|\left|\frac{\mathbf{q}(s,\mathbf{d},\boldsymbol{\tau})}{\psi_0(s,\mathbf{d}^0)}\right| \ge \left|\psi_0(s,\mathbf{d}^0)\right|\left[\left|\frac{\psi_0(s,\mathbf{d})}{\psi_0(s,\mathbf{d}^0)}\right| - \sum_{k=1}^{K}\left|\frac{\psi_k(s,\mathbf{d})}{\psi_0(s,\mathbf{d}^0)}\right|\right] > 0,$$

and $|\mathbf{q}(s,\mathbf{d},\boldsymbol{\tau})| > 0$. To complete the proof we have to show the existence of $\epsilon > 0$ so that quasipolynomials (6.1.8) do not vanish in the bounded region

$$\{s \;:\; \text{Re } s \ge 0, \;\text{ and } |s| \le \rho\}\,.$$

The proof of the lemma now follows from the continuous dependence of $\mathbf{q}(s,\mathbf{d},\boldsymbol{\tau})$ on $(\mathbf{d},\boldsymbol{\tau})$.

We return now to the quasipolynomial segment (6.1.7).

Lemma 6.1.2 *Suppose that $\mathbf{q}\left(s,\mathbf{d}^\lambda,\boldsymbol{\tau}^\lambda\right)$ is an unstable quasipolynomial. If there exists s_λ such that*

1. $\mathbf{q}\left(s_\lambda,\mathbf{d}^\lambda,\boldsymbol{\tau}^\lambda\right) = 0,$
2. Re $s_\lambda > 0,$

then for each $\epsilon > 0$ there exists λ^*, $\lambda - \epsilon < \lambda^* < \lambda$ such that

1. $\mathbf{q}\left(s_{\lambda^*},\mathbf{d}^{\lambda^*},\boldsymbol{\tau}^{\lambda^*}\right) = 0,$
2. Re $s_{\lambda^*} > 0.$

Proof: See Theorem of Hurwitz [Ti].

Theorem 6.1.3 If $\mathbf{q}(s,\mathbf{d}^0,\boldsymbol{\tau}^0)$ is a stable quasipolynomial, and $\mathbf{q}(s,\mathbf{d}^1,\boldsymbol{\tau}^1)$ is an unstable quasipolynomial, then there exists $\lambda_0 \in [0,1]$, and s_{λ_0} such that

1. $\mathbf{q}\left(s_{\lambda_0},\mathbf{d}^{\lambda_0},\boldsymbol{\tau}^{\lambda_0}\right) = 0,$
2. Re $s_{\lambda_0} = 0.$

Proof: Let $\lambda_0 = \inf\left\{\lambda \;:\; 0 \le \lambda \le 1,\; q\left(s,\mathbf{d}^\lambda,\boldsymbol{\tau}^\lambda\right) \text{ is unstable}\right\}$. According to Lemma 6.1.1 the quasipolynomial $\mathbf{q}\left(s,\mathbf{d}^{\lambda_0},\boldsymbol{\tau}^{\lambda_0}\right)$ is unstable. Due to Assumption 6.1.1 the unstable roots of $\mathbf{q}\left(s,\mathbf{d}^{\lambda_0},\boldsymbol{\tau}^{\lambda_0}\right)$ are bounded. According to Lemma 6.1.2 there exists an unstable root s_{λ_0} with Re $s_{\lambda_0} = 0$. The proof of the theorem is now completed.

6.2. Real Interval Quasipolynomials

In the following sections we consider a quasipolynomial family with interval coefficients and interval delays

$$\mathbf{Q} = \left\{ \mathbf{q}(s, \mathbf{a}, \boldsymbol{\tau}) = \sum_{n,k=0}^{N,K} a_{n,k} s^n e^{-\tau_k s} \right\}. \tag{6.2.1}$$

The interval coefficients $\mathbf{a} = (a_{0,0}, \ldots, a_{N,K})^t$ satisfy

$$\underline{a_{n,k}} \leq a_{n,k} \leq \overline{a_{n,k}}, \qquad n = 0, 1, \ldots, N, \ k = 0, 1, \ldots, K, \tag{6.2.2}$$

and the delays $\boldsymbol{\tau} = (\tau_0, \tau_1, \ldots, \tau_K)^t$ vary in the intervals

$$0 = \tau_0, \qquad 0 < \underline{\tau}_k \leq \tau_k \leq \overline{\tau}_k, \qquad k = 1, \ldots, K. \tag{6.2.3}$$

Since

$$0 = \mathbf{q}\left(jw, \mathbf{a}^{\lambda_0}, \boldsymbol{\tau}^{\lambda_0}\right) = \overline{\mathbf{q}\left(jw, \mathbf{a}^{\lambda_0}, \boldsymbol{\tau}^{\lambda_0}\right)} = \mathbf{q}\left(-jw, \mathbf{a}^{\lambda_0}, \boldsymbol{\tau}^{\lambda_0}\right),$$

and the value set $\mathcal{Q}_0$ is the real interval $\left[\sum_{k=0}^{K} \underline{a_{0,k}}, \sum_{k=0}^{K} \overline{a_{0,k}}\right]$, the necessary and sufficient robust stability conditions given by Theorem 6.1.2 become the following.

Theorem 6.2.1 Suppose that $0 \notin \mathcal{Q}_0$. The quasipolynomial family $\mathbf{Q}$ is stable if and only if

$$0 \notin \partial \mathcal{Q}_w \text{ for each } w > 0. \tag{6.2.4}$$

6.3. Kharitonov Rectangles and Robust Stability Conditions

We start with a useful observation due to Chapellat and Bhattacharyya [CB] (see also Remark 4.5.3, page 86). When $s = jw$ the polynomials $\psi_k(s, \mathbf{a})$ can be written as

$$\psi_k(s, \mathbf{a}) = \mathbf{p}_k(s) + x_k \mathbf{p}_{xk}(s) + j y_k \mathbf{p}_{yk}(s),$$

where $0 \leq x_k, \ y_k \leq 1, \ k = 0, \ldots, K$; and .

$$\begin{aligned}
\mathbf{p}_k(s) &= \underline{a_{0,k}} + \underline{a_{1,k}} s + \overline{a_{2,k}} s^2 + \quad \ldots \\
\mathbf{p}_{xk}(s) &= \left[\overline{a_{0,k}} - \underline{a_{0,k}}\right] + \left[\underline{a_{2,k}} - \overline{a_{2,k}}\right] s^2 + \quad \ldots \\
j\mathbf{p}_{yk}(s) &= \left[\overline{a_{1,k}} - \underline{a_{1,k}}\right] s + \left[\underline{a_{3,k}} - \overline{a_{3,k}}\right] s^3 + \quad \ldots
\end{aligned} \tag{6.3.1}$$

Since the intervals for odd and even coefficients are independent the value set $\mathcal{P}_w^k$ of the polynomial family $\psi_k(s, \mathbf{a})$ is a rectangle $B_k(w)$ with the vertices generated by four Kharitonov polynomials.

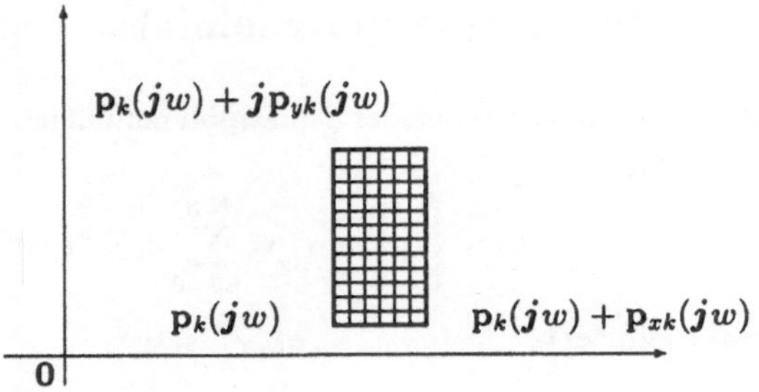

The value set $\mathcal{Q}_w$ of the quasipolynomial family (6.2.1) will be written in the following form

$$\begin{aligned} \mathcal{Q}_w &= \left\{ \sum_{k=0}^{K} \left[\mathbf{p}_k(jw) + x_k \mathbf{p}_{xk}(jw) + j y_k \mathbf{p}_{yk}(jw) \right] e^{-\tau_k j w} \right\} \\ &= \left\{ B_0(w) + \sum_{k=1}^{K} B_k(w) e^{-\tau_k j w} \right\}. \end{aligned}$$

To simplify the exposition we assume throughout that

$$\mathbf{p}_{xk}(jw) > 0, \text{ and } \mathbf{p}_{yk}(jw) > 0 \text{ for each } w > 0, \text{ and } k = 0, \ldots, K.$$

Consider the $3K+2$ dimensional box

$$\mathbf{B} = \left\{ (\mathbf{x}, \mathbf{y}, \boldsymbol{\tau}) \ : \ \begin{array}{ll} 0 \le x_k, \ y_k \le 1 & k = 0, 1, \ldots, K \\ \underline{\tau}_k \le \tau_k \le \overline{\tau}_k & k = 1, \ldots, K \end{array} \right\}. \tag{6.3.2}$$

For $w > 0$ the value set $\mathcal{Q}_w$ is the image of the box under the mapping $f \ : \ \mathbf{B} \to \mathbf{C}$ defined by

$$f(\mathbf{x}, \mathbf{y}, \boldsymbol{\tau}) = \sum_{k=0}^{K} \left[\mathbf{p}_k(jw) + x_k \mathbf{p}_{xk}(jw) + j y_k \mathbf{p}_{yk}(jw) \right] e^{-\tau_k j w}. \tag{6.3.3}$$

The set of one dimensional edges of the box $\mathcal{E}$ and a family of principal segments to be introduced later are crucial for robust stability of the quasipolynomial family $\mathbf{Q}$.

Theorem 6.3.1 At each $w > 0$ there exists a finite set of principal segments $\mathbf{S}(w)$ of $\mathbf{B}$ such that $\partial f(\mathbf{B}) \subseteq f\left(\mathcal{E} \bigcup \mathbf{S}(w)\right)$.

The proof of the theorem is given in Section 6.4. The formulae for $\mathbf{S}(w)$ are given in Section 6.5. An example of a family of the principal segments is given below.

$$x_1 = \ldots = x_K = y_1 = \ldots = y_K = 0, \ \underline{\tau}_1 \le \tau_1 \le \overline{\tau}_1, \ \tau_3 = \underline{\tau}_3, \ldots, \tau_K = \underline{\tau}_K,$$

$$\tau_2 = \tau_1 + \frac{1}{w} \left[\arg \frac{\mathbf{p}_2(jw)}{\mathbf{p}_1(jw)} + \pi n \right], \ n = 0, \pm 1, \pm 2, \ldots \tag{6.3.4}$$

Since $\tau_2 \in [\underline{\tau}_2, \overline{\tau}_2]$ relations (6.3.4) defines a finite set of linear segments. Necessary and sufficient robust stability conditions are given next.

Theorem 6.3.2 The quasipolynomial family **Q** given by (6.2.1) is stable if and only if

1. $\left[\sum_{k=0}^{K} \underline{a_{0,k}}\right]\left[\sum_{k=0}^{K} \overline{a_{0,k}}\right] > 0.$

2. For each $w > 0$ one has $0 \notin f\left(\mathcal{E} \bigcup \mathbf{S}(w)\right)$.

Notice, that the number of the segments defined by (6.3.4) increases to infinity as $w \to \infty$. A significant computational effort is needed to verify the last condition of the theorem for large w. To overcome this difficulty we consider the set

$$\left\{B_k(w)e^{-\tau_k \jmath w} \ : \ \underline{\tau}_k \leq \tau_k \leq \overline{\tau}_k\right\}.$$

When $w \geq w_k = \dfrac{2\pi}{\overline{\tau}_k - \underline{\tau}_k}$ the set is an annulus $A(\underline{r}_k(w), \overline{r}_k(w))$, where $\underline{r}_k(w)$ and $\overline{r}_k(w)$ are the inner and outer radii of the annulus. If $w_* = \max\{w_1, \ldots, w_K\}$, and $w \geq w_*$ the value set is given by

$$\mathcal{Q}_w = B_0(w) + \sum_{k=1}^{K} A(\underline{r}_k(w), \overline{r}_k(w)). \tag{6.3.5}$$

The last term in (6.3.5) is an annulus $A(\underline{r}(w), \overline{r}(w))$ whose inner and outer radii $\underline{r}(w)$ and $\overline{r}(w)$ can be easily evaluated, in particular $\overline{r}(w) = \sum_{k=1}^{K} \overline{r}_k(w)$. For $w \geq w_*$ the last condition of Theorem 6.3.2 can be substituted by

$$B_0(w) \bigcap A(\underline{r}(w), \overline{r}(w)) = \emptyset. \tag{6.3.6}$$

Furthermore, due to Assumption 6.1.1 for large w relation (6.3.6) holds if and only if

$$B_0(w) \bigcap D(\overline{r}(w)) = \emptyset,$$

where $D(\overline{r}(w))$ stands for the two dimensional disc of radius $\overline{r}(w)$ centered at the origin. The verification of this condition is straightforward. Numerically implementable necessary and sufficient robust stability conditions are given next.

Theorem 6.3.3 The quasipolynomial family **Q** given by (6.2.1) is stable if and only if

1. $\left[\sum_{k=0}^{K} \underline{a_{0,k}}\right]\left[\sum_{k=0}^{K} \overline{a_{0,k}}\right] > 0.$

2. For each $0 < w < w_*$ one has $0 \notin f\left(\mathcal{E} \bigcup \mathbf{S}(w)\right)$.

3. For each $w \geq w_*$ one has $B_0(w) \bigcap D(\overline{r}(w)) = \emptyset$.

Furthermore, the degree condition of Assumption 6.1.1 yields the existence of w^* such that $B_0(w) \bigcap D(\overline{r}(w)) = \emptyset$ when $w^* < w$. If $w_* < w^*$, then the last condition of the theorem should be checked over the finite interval $[w_*, w^*]$ only.

6.4. Image of a Box Under Nonlinear Mapping

The proof of Theorem 6.3.1 is based on degree theory results presented below.

Let $\mathbf{O}$ be an open bounded subset of $\mathbf{R}^n$, and ϕ is a continuous function from $\overline{\mathbf{O}}$ into $\mathbf{R}^n$ with the norm

$$|\phi|_\infty = \sup\{|\phi(\mathbf{x})|_\infty \ : \ \mathbf{x} \in \mathbf{O}\}.$$

With each $\mathbf{y} \notin \phi(\partial\mathbf{O})$ we associate a number $d(\phi, \mathbf{O}, \mathbf{y})$ *the degree of* ϕ *at* $\mathbf{y}$ *relative to* $\mathbf{O}$. The degree enjoys the following properties that will be useful for the proof of Lemma 6.4.1.

1. Let ψ be the identity mapping, i.e., $\psi(\mathbf{x}) = \mathbf{x}$ for each $\mathbf{x} \in \overline{\mathbf{O}}$. If $\mathbf{y} \in \mathbf{O}$, then $d(\psi, \mathbf{O}, \mathbf{y}) = 1$; if $\mathbf{y} \notin \overline{\mathbf{O}}$, then $d(\psi, \mathbf{O}, \mathbf{y}) = 0$.
2. If $d(\phi, \mathbf{O}, \mathbf{y})$ is defined and nonzero, then there exists $\mathbf{x} \in \mathbf{O}$ such that $\phi(\mathbf{x}) = \mathbf{y}$.
3. Suppose that $\mathbf{y} \notin \psi(\partial\mathbf{O})$. If $|\psi - \phi|_\infty < \inf\{|\mathbf{y} - \psi(\mathbf{x}_\partial)|_\infty \ : \ \mathbf{x}_\partial \in \partial\mathbf{O}\}$, then $d(\phi, \mathbf{O}, \mathbf{y})$ is defined and equal to $d(\psi, \mathbf{O}, \mathbf{y})$.

For formal definitions and proofs we refer the reader to [Ll].

To simplify the presentation we consider a continuously differentiable mapping

$$f \ : \ \mathbf{B}^m \to \mathbf{C}$$

from the unit box $\mathbf{B}^m = \{\mathbf{x} \ : \ 0 \le x_i \le 1\}$ to the complex plane. Denote the partial derivatives of f by $G_1(\mathbf{x}), \ldots, G_m(\mathbf{x})$. Let $D_i(\mathbf{x}), i = 1, 2, \ldots, 2m$ be complex numbers defined by (4.1.11), i.e.,

$$\begin{array}{llll} D_i(\mathbf{x}) = G_i(\mathbf{x}) & \text{and} \ \ D_{i+m}(\mathbf{x}) = 0 & \text{if} & 0 = x_i, \\ D_i(\mathbf{x}) = G_i(\mathbf{x}) & \text{and} \ \ D_{i+m}(\mathbf{x}) = -G_i(\mathbf{x}) & \text{if} & 0 < x_i < 1, \\ D_i(\mathbf{x}) = 0 & \text{and} \ \ D_{i+m}(\mathbf{x}) = -G_i(\mathbf{x}) & \text{if} & x_i = 1. \end{array} \tag{6.4.1}$$

Lemma 6.4.1 *For* $\mathbf{x} \in \mathbf{B}^m$ *let*

$$D_{i_1} \prec D_{i_2} \prec \ldots \prec D_{i_k} \tag{6.4.2}$$

be the list of all distinct nonzero complex numbers associated with $\mathbf{x}$ *written in the order of increasing argument. If the list is not empty, and*

$$arg\,\frac{D_{i_2}}{D_{i_1}} < \pi, \ \ldots, \ arg\,\frac{D_{i_1}}{D_{i_k}} < \pi, \tag{6.4.3}$$

then there exists a square $B \in \mathbf{C}$ *centered at* $f(\mathbf{x})$ *and covered by* $f(\mathbf{B}^m)$, *and* $f(\mathbf{x})$ *is an interior point of* $f(\mathbf{B}^m)$.

Proof: Relations (6.4.3) imply that each $z \in \mathbf{C}$ belongs to a convex cone generated by the origin, D_{i_j}, and $D_{i_{j+1}}$ for some $j \in \{1, \dots, k\}$. That is

$$z = \gamma_1 D_{i_j} + \gamma_2 D_{i_{j+1}}, \; 0 \le \gamma_1, \; \gamma_2$$

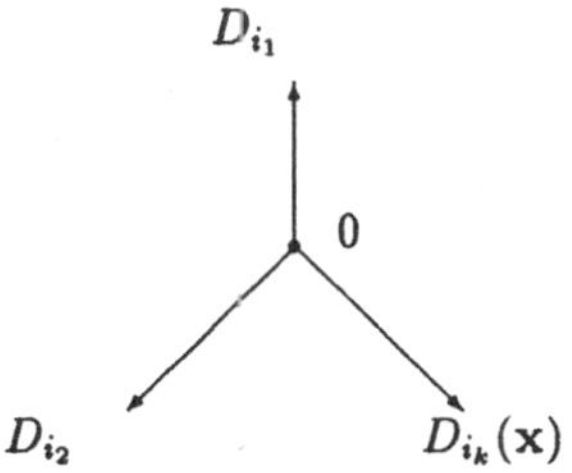

Let ϵ_1 and ϵ_2 be real numbers whose signs are defined as follows

$$\begin{array}{llllll} 0 \le \epsilon_1 & \text{if} \quad 1 \le i_j \le m, & \text{and} & \epsilon_1 \le 0 & \text{if} & m+1 \le i_j \le 2m, \\ 0 \le \epsilon_2 & \text{if} \quad 1 \le i_{j+1} \le m, & \text{and} & \epsilon_2 \le 0 & \text{if} & m+1 \le i_{j+1} \le 2m. \end{array} \tag{6.4.4}$$

The mapping $\phi \; : \; \mathbf{C} \rightarrow \mathbf{C}$ introduced below is instrumental for the proof of the lemma. For $z = \gamma_1 D_{i_j} + \gamma_2 D_{i_{j+1}}$ define

$$\phi(z) = f(x_1, \dots, x_{l_j - 1}, x_{l_j} + \epsilon_1, x_{l_j + 1}, \dots, x_{l_{j+1} - 1}, x_{l_{j+1}} + \epsilon_2, x_{l_{j+1}+1}, \dots, x_m) \tag{6.4.5}$$

where

$$\begin{array}{lll} l_j = i_j & \text{if} & 1 \le i_j \le m, \\ & & \qquad |\epsilon_1| = \gamma_1, \; |\epsilon_2| = \gamma_2, \\ l_j = i_j - m & \text{if} & m+1 \le i_j \le 2m, \end{array}$$

and the signs of ϵ_1 and ϵ_2 are determined by (6.4.4). In the sequel we denote

$$(x_1, \dots, x_{l_j - 1}, x_{l_j} + \epsilon_1, x_{l_j+1}, \dots, x_{l_{j+1}-1}, x_{l_{j+1}} + \epsilon_2, x_{l_{j+1}+1}, \dots, x_m)^t \text{ by } \mathbf{x}_{l_j l_{j+1}}(\epsilon_1, \epsilon_2).$$

Notice that $\mathbf{x}_{l_j l_{j+1}}(\epsilon_1, \epsilon_2) \in \mathbf{B}^m$ when $|\epsilon_1|$ and $|\epsilon_2|$ are small. Furthermore,

$$f(\mathbf{x}_{l_j l_{j+1}}(\epsilon_1, \epsilon_2)) - f(\mathbf{x}) = |\epsilon_1| \, D_{i_j} + |\epsilon_2| \, D_{i_{j+1}} + o(\boldsymbol{\epsilon}), \tag{6.4.6}$$

where $\boldsymbol{\epsilon} = (\epsilon_1, \epsilon_2)$, and $\lim\limits_{|\boldsymbol{\epsilon}|_\infty \to 0} \dfrac{o(\boldsymbol{\epsilon})}{|\boldsymbol{\epsilon}|_\infty} = 0$. Consider a convex polygon P in the complex plane whose vertices are

$$\{f(\mathbf{x}) + D_{i_1}, \; \dots, f(\mathbf{x}) + D_{i_k}\}.$$

The boundary ∂P of the polygon is made up of the k segments

$$\left\{\gamma \left[D_{i_j} + f(\mathbf{x})\right] + (1-\gamma)\left[D_{i_{j+1}} + f(\mathbf{x})\right] \; : \; 0 \le \gamma \le 1\right\}.$$

Due to (6.4.3) the distance between ∂P and the origin is a positive number which we denote by r. Let δ_j, $j = 1, \ldots, k$ be a positive number such that

$$\mathbf{x}_{l_j l_{j+1}}(\epsilon_1, \epsilon_2) \in \mathbf{B}^m \text{ and } \frac{o(\epsilon)}{|\epsilon|_\infty} < \frac{r}{4}, \quad \text{where} \quad |\epsilon|_\infty < \delta_j. \tag{6.4.7}$$

Finally let $\delta = \min\{\delta_1, \ldots, \delta_k\}$, and consider the convex polygon P_δ in the complex plane whose vertices are

$$\{f(\mathbf{x}) + \delta D_{i_1}, \ \ldots, f(\mathbf{x}) + \delta D_{i_k}\}$$

and the square B_δ centered at $f(\mathbf{x})$,

$$B_\delta = \left\{z \ : \ |z - f(\mathbf{x})|_\infty \le \frac{r\delta}{4}\right\}.$$

To complete the proof we next show that $B_\delta \subseteq f(\mathbf{B}^m)$. Notice that for each $z_B \in B_\delta$, and $z_\partial \in \partial P_\delta$ one has

$$|z_B - z_\partial|_\infty = |[z_B - f(\mathbf{x})] + [f(\mathbf{x}) - z_\partial]|_\infty \ge r\delta - \frac{r\delta}{4} = \frac{3}{4}r\delta. \tag{6.4.8}$$

Furthermore for each choice of

$$z = |\epsilon_1|\, D_{i_j} + |\epsilon_2|\, D_{i_j+1} \in P_\delta, \ z_B \in B_\delta, \text{ and } z_\partial \in \partial P_\delta$$

one has

$$|z - \phi(z)|_\infty \le \frac{1}{4}r\, |\epsilon|_\infty \le \frac{1}{4}r\delta < \frac{3}{4}r\delta < |z_B - z_\partial|_\infty. \tag{6.4.9}$$

Inequality (6.4.9) shows that

$$|\psi - \phi|_\infty < \inf\{|z_B - z_\partial|_\infty \ : \ z_\partial \in \partial P_\delta\},$$

for $\psi(z) \equiv z$. This yields $d(\phi, \mathbf{O}, z_B) = d(\psi, \mathbf{O}, z_B) = 1$, where $\mathbf{O}$ is the interior of P_δ. Hence there exists $z = |\epsilon_1| D_{i_j} + |\epsilon_2| D_{i_{j+1}} \in P_\delta$ such that $z_B = \phi(z) = f(\mathbf{x}_{l_j l_{j+1}}(\epsilon_1, \epsilon_2))$, i.e., $z_B \in f(\mathbf{B}^m)$.

Lemma 6.4.2 *Let* $\mathbf{x} \in \mathbf{B}^m$. *If* $f(\mathbf{x}) \in \partial f(\mathbf{B}^m)$, *then there exists a nonzero complex number* G *such that*

$$D_i(\mathbf{x}) \preceq G \text{ for each nonzero } D_i(\mathbf{x}). \tag{6.4.10}$$

Proof: Let $D_{i_1} \prec D_{i_2} \prec \ldots \prec D_{i_k}$ be the list of all distinct nonzero complex numbers associated with $\mathbf{x}$ written in the order of increasing argument. Assume first that $k \ge 3$. According to Lemma 6.4.1 there exists an index j such that $\arg \dfrac{D_{i_j}}{D_{i_{j+1}}} \ge \pi$. Hence $G = D_{i_j}$ satisfies (6.4.10). The case $k \le 2$ is trivial.

Finally notice that a point $\mathbf{x}$ that satisfies (6.4.10) is a principal point. The proof of Theorem 6.3.1 is now completed.

6.5. Principal Segments

The main result of the section is Lemma 6.5.4 that shows that the set of principal segments $\mathbf{S}(w)$ is a finite collection of linear segments of the box $\mathbf{B}$ defined by (6.3.2).

The partial derivatives of the function $f(\mathbf{x},\mathbf{y},\boldsymbol{\tau})$ given by (6.3.3) are denoted as follows

$$\frac{\partial f}{\partial x_i} = G_{xi}, \ \frac{\partial f}{\partial y_i} = G_{yi}, \ \frac{\partial f}{\partial \tau_i} = G_{\tau i}.$$

To simplify the presentation we assume that all the partial derivatives do not vanish.

Lemma 6.5.1 *Suppose that $f(\mathbf{x},\mathbf{y},\boldsymbol{\tau}) \in \partial f(\mathbf{B})$. If x_i is a free coordinate, then y_i is an extremal coordinate. If y_i is a free coordinate, then x_i is an extremal coordinate.*

Proof: Suppose that both x_i and y_i are free coordinates. Then

$$D_{xi} = G_{xi} = \mathbf{p}_{xi}(\boldsymbol{j}w)e^{-\tau_i \boldsymbol{j} w} \asymp e^{-\tau_i \boldsymbol{j} w} \neq 0, \qquad D_{x(i+K+1)} = -D_{xi} \neq 0,$$
$$D_{yi} = G_{yi} = \boldsymbol{j}\mathbf{p}_{yi}(\boldsymbol{j}w)e^{-\tau_i \boldsymbol{j} w} \asymp \boldsymbol{j}e^{-\tau_i \boldsymbol{j} w} \neq 0, \qquad D_{y(i+K+1)} = -D_{yi} \neq 0,$$

where, for example,

$$D_{xi}(\mathbf{x},\mathbf{y},\boldsymbol{\tau}) = G_{xi}(\mathbf{x},\mathbf{y},\boldsymbol{\tau}) \quad \text{and} \quad D_{x(i+K+1)}(\mathbf{x},\mathbf{y},\boldsymbol{\tau}) = -G_{xi}(\mathbf{x},\mathbf{y},\boldsymbol{\tau}).$$

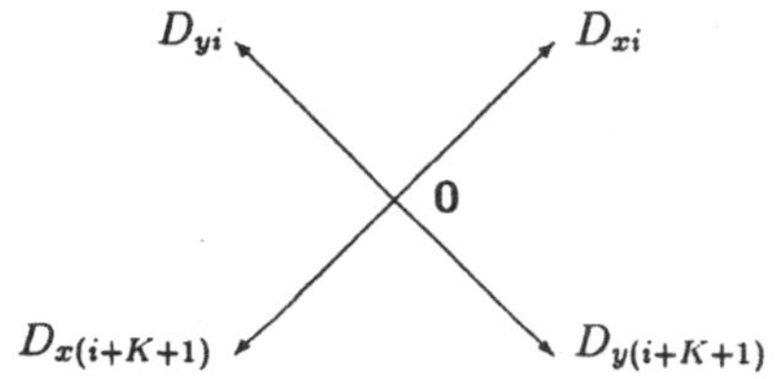

This contradicts Lemma 6.4.2, and completes the proof.

Lemma 6.5.2 *Suppose that $f(\mathbf{x},\mathbf{y},\boldsymbol{\tau}) \in \partial f(\mathbf{B})$, and τ_i is a free coordinate. If x_i is a free coordinate, then $x_i = -\dfrac{\text{Re}\ \mathbf{p}_i(\boldsymbol{j}w)}{\mathbf{p}_{xi}(\boldsymbol{j}w)}$; if y_i is a free coordinate, then $y_i = -\dfrac{\text{Im}\ \mathbf{p}_i(\boldsymbol{j}w)}{\mathbf{p}_{yi}(\boldsymbol{j}w)}$.*

Proof: Notice that

$$G_{\tau i} \asymp \boldsymbol{j}\left[\mathbf{p}_i(\boldsymbol{j}w) + x_i\mathbf{p}_{xi}(\boldsymbol{j}w) + \boldsymbol{j}y_i\mathbf{p}_{yi}(\boldsymbol{j}w)\right] e^{-\tau_i \boldsymbol{j} w}, \ G_{xi} \asymp e^{-\tau_i \boldsymbol{j} w}, \ G_{yi} \asymp \boldsymbol{j}e^{-\tau_i \boldsymbol{j} w}.$$

If x_i is a free coordinate, then $G_{\tau i} \asymp G_{xi}$, i.e., $\mathbf{Re}\ \mathbf{p}_i(\boldsymbol{j}w) + x_i\mathbf{p}_{xi}(\boldsymbol{j}w) = 0$, and $x_i = -\dfrac{\text{Re}\ \mathbf{p}_i(\boldsymbol{j}w)}{\mathbf{p}_{xi}(\boldsymbol{j}w)}$. The proof of the second part of the lemma is analogous.

Definition 6.5.1 *For $z \in \partial f(\mathbf{B})$ let $(\mathbf{x}^z, \mathbf{y}^z, \boldsymbol{\tau}^z)$ be an element of* **B** *such that*

1. $f(\mathbf{x}^z, \mathbf{y}^z, \boldsymbol{\tau}^z) = z$,

2. *for each* $(\mathbf{x}, \mathbf{y}, \boldsymbol{\tau}) \in \mathbf{B}$ *with* $f(\mathbf{x}, \mathbf{y}, \boldsymbol{\tau}) = z$ *one has* $|I_f(\mathbf{x}^z, \mathbf{y}^z, \boldsymbol{\tau}^z)| \leq |I_f(\mathbf{x}, \mathbf{y}, \boldsymbol{\tau})|$.

Lemma 6.5.3 *If $z \in \partial f(\mathbf{B})$, then $|I_f(\mathbf{x}^z, \mathbf{y}^z)| \leq 1$, and $G_{\tau i} \neq 0$ for each $i \in I_f(\boldsymbol{\tau}^z)$.*

Proof: If, for example, $|I_f(\mathbf{x}^z, \mathbf{y}^z)| \geq 2$, then there exists two indices $i \neq j$ such that one of the following four conditions holds

$$\begin{array}{lcll} x_i^z & \text{and} & x_j^z & \text{are free coordinates} \\ x_i^z & \text{and} & y_j^z & \text{are free coordinates} \\ y_i^z & \text{and} & x_j^z & \text{are free coordinates} \\ y_i^z & \text{and} & x_j^z & \text{are free coordinates} \end{array}$$

If x_i^z and x_j^z are free coordinates, then $G_{xi} = cG_{xk}$. Hence for each real γ

$$f(x_1^z, \ldots, x_i^z + \gamma, \ldots, x_j^z - c\gamma, \ldots, x_K^z, \mathbf{y}^z, \boldsymbol{\tau}^z) = f(\mathbf{x}^z, \mathbf{y}^z, \boldsymbol{\tau}^z).$$

There exists γ so that the point $\left(x_i^z + \gamma, x_j^z - c\gamma\right)$ hits the boundary of the two dimensional square $[0,1] \times [0,1]$. That is

$$|I_f(\mathbf{x}, \mathbf{y}^z, \boldsymbol{\tau}^z)| < |I_f(\mathbf{x}^z, \mathbf{y}^z, \boldsymbol{\tau}^z)|, \text{ where } \mathbf{x} = (x_1^z, \ldots, x_i^z + \gamma, \ldots, x_j^z - c\gamma, \ldots, x_K^z)^t.$$

This contradicts Definition 6.5.1, and completes the proof of the first case. The proofs of the reminding cases are analogous. To complete the proof of the lemma we notice that $G_{\tau i} = 0$ implies

$$\mathbf{p}_i(jw) + x_i \mathbf{p}_{xi}(jw) + j y_i \mathbf{p}_{yi}(jw) = 0,$$

and shows that for $\boldsymbol{\tau} = \left(\tau_1^z, \tau_2^z, \ldots, \tau_{i-1}^z, \underline{\tau}_i, \tau_{i+1}^z, \ldots, \tau_K^z\right)^t$ one has

$$f(\mathbf{x}^z, \mathbf{y}^z, \boldsymbol{\tau}) = f(\mathbf{x}^z, \mathbf{y}^z, \boldsymbol{\tau}^z), \text{ and } |I_f(\mathbf{x}^z, \mathbf{y}^z, \boldsymbol{\tau})| < |I_f(\mathbf{x}^z, \mathbf{y}^z, \boldsymbol{\tau}^z)|.$$

This contradiction completes the proof of the lemma.

Lemma 6.5.4 *If $z \in \partial f(\mathbf{B})$, and $|I_f(\mathbf{x}^z, \mathbf{y}^z, \boldsymbol{\tau}^z)| \geq 2$, then one of the following five cases holds for $n = 0, \pm 1, \ldots$:*

Case 1. *If $|I_f(\mathbf{x}^z, \mathbf{y}^z)| = 0$, and $\tau_{i_1}^z, \ldots, \tau_{i_k}^z$ are free coordinates of $(\mathbf{x}^z, \mathbf{y}^z, \boldsymbol{\tau}^z)$, then for each $j = 2, \ldots, k$*

$$\tau_{i_j}^z = \tau_{i_1}^z + \frac{1}{w}\left[arg \frac{\mathbf{p}_{i_j}(jw) + x_{i_j}^z \mathbf{p}_{xi_j}(jw) + j y_{i_j}^z \mathbf{p}_{yi_j}(jw)}{\mathbf{p}_{i_1}(jw) + x_{i_1}^z \mathbf{p}_{xi_1}(jw) + j y_{i_1}^z \mathbf{p}_{yi_1}(jw)} + \pi n\right]. \qquad (6.5.1)$$

Case 2. *If* $|I_f(\mathbf{x}^z, \mathbf{y}^z)| = 1$, x_i^z, *and,* $\tau_{i_1}^z, \ldots, \tau_{i_k}^z$ *are free coordinates, and* $i \notin \{i_1, \ldots, i_k\}$, *then (6.5.1) holds, and in addition,*

$$\tau_{i_1}^z = \frac{1}{w}\left(\frac{\pi}{2} + \pi n + arg\left[\mathbf{p}_{i_1}(jw) + x_{i_1}^z \mathbf{p}_{xi_1}(jw) + jy_{i_1}^z \mathbf{p}_{yi_1}(jw)\right]\right) + \tau_i^z. \qquad (6.5.2)$$

Case 3. *If* $|I_f(\mathbf{x}^z, \mathbf{y}^z)| = 1$, y_i^z, *and,* $\tau_{i_1}^z, \ldots, \tau_{i_k}^z$ *are free coordinates, and* $i \notin \{i_1, \ldots, i_k\}$, *then (6.5.1) holds, and in addition,*

$$\tau_{i_1}^z = \frac{1}{w}\left(\pi n + arg\left[\mathbf{p}_{i_1}(jw) + x_{i_1}^z \mathbf{p}_{xi_1}(jw) + jy_{i_1}^z \mathbf{p}_{yi_1}(jw)\right]\right) + \tau_i^z. \qquad (6.5.3)$$

Case 4. *If* $|I_f(\mathbf{x}^z, \mathbf{y}^z)| = 1$, $x_{i_1}^z$, *and* $\tau_{i_1}^z, \ldots, \tau_{i_k}^z$ *are free coordinates, then (6.5.1) holds, and*

$$x_{i_1}^z = -\frac{\text{Re } \mathbf{p}_{i_1}(jw)}{\mathbf{p}_{xi_1}(jw)}.$$

Case 5. *If* $|I_f(\mathbf{x}^z, \mathbf{y}^z)| = 1$, $y_{i_1}^z$, *and* $\tau_{i_1}^z, \ldots, \tau_{i_k}^z$ *are free coordinates, then (6.5.1) holds, and*

$$y_{i_1}^z = -\frac{\text{Im } \mathbf{p}_{i_1}(jw)}{\mathbf{p}_{yi_1}(jw)}.$$

The points $(\mathbf{x}^z, \mathbf{y}^z, \boldsymbol{\tau}^z)$ *described by the five cases are denoted by* $\mathbf{S}(w)$. *The set* $\mathbf{S}(w)$ *contains a finite number of one dimensional linear segments of the box* $\mathbf{B}$.

Proof: If $|I_f(\mathbf{x}^z, \mathbf{y}^z)| = 0$, and $\tau_{i_1}^z, \ldots, \tau_{i_k}^z$ are free coordinates of $(\mathbf{x}^z, \mathbf{y}^z, \boldsymbol{\tau}^z)$, then equations (6.5.1) follows straightforward from the relations

$$G_{\tau i_1} \asymp G_{\tau i_j}, \; j = 2, \ldots, k.$$

The proofs for the other cases are analogous.

We now complete the proof of Theorem 6.3.1. Let $z \in \partial f(\mathbf{B})$. If $|I_f(\mathbf{x}^z, \mathbf{y}^z, \boldsymbol{\tau}^z)| \leq 1$, then the point $(\mathbf{x}^z, \mathbf{y}^z, \boldsymbol{\tau}^z)$ belongs to a one dimensional edge of $\mathbf{B}$. Otherwise $(\mathbf{x}^z, \mathbf{y}^z, \boldsymbol{\tau}^z) \in \mathbf{S}(w)$. Hence $\partial f(\mathbf{B}) \subseteq f\left(\mathcal{E} \bigcup \mathbf{S}(w)\right)$. The proof of Theorem 6.3.1 is now completed.

6.6. Numerical Example

Fu, Olbrot and Polis [FOP1] show that the quasipolynomial $\gamma + s^2 + 2se^{-s} + e^{-2s}$ is stable for each $0 \leq \gamma \leq \left(\frac{\pi}{2} - 1\right)^2 \approx 0.3258$. Motivated by this result we consider the quasipolynomial family with two interval delays:

$$\mathbf{Q} = \left\{\left[a_{0,0} + a_{1,0}s + s^2\right] + 2se^{-\tau_1 s} + e^{-\tau_2 s}\right\}, \qquad (6.6.1)$$

where

$$|a_{0,0}| \leq \gamma, \ |a_{1,0}| \leq \gamma, \ |1 - \tau_1| \leq \gamma, \ |2 - \tau_2| \leq \gamma, \quad \text{and } 0 \leq \gamma < 1.$$

Notice that the nominal quasipolynomial $\left(s + e^{-s}\right)^2$ is stable (see e.g., [BC]). The value set $\mathcal{Q}_0 = [-\gamma + 1, \gamma + 1]$ does not contain the origin for each $0 \leq \gamma < 1$. For the family (6.6.1) one has $w_* = \frac{\pi}{\gamma}$. When $w \geq w_*$ the vertices of the rectangle $B_0(w)$ are

$$\left(-\gamma - w^2, -j\gamma w\right), \ \left(\gamma - w^2, -j\gamma w\right), \ \left(\gamma - w^2, j\gamma w\right), \ \left(-\gamma - w^2, j\gamma w\right),$$

and $\overline{r}(w) = 2w + 1$.

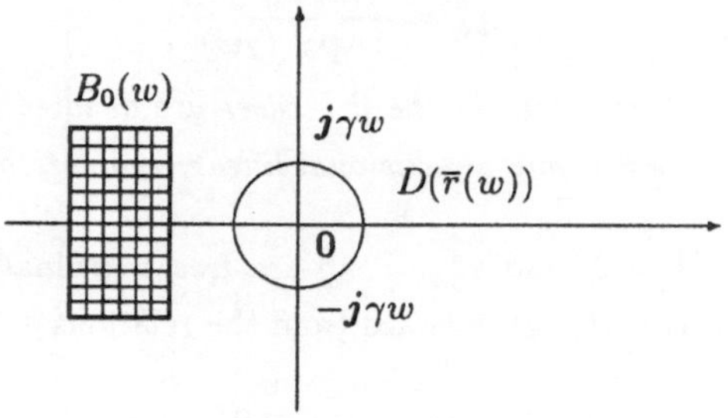

When $w \geq w_*$ the distance from $B_0(w)$ to the origin is $w^2 - \gamma$, and a straightforward computation shows that

$$w^2 - \gamma > 2w + 1 = \overline{r}(w) \text{ for each } w \geq w_*.$$

To verify stability of the family on has to check the condition

$$0 \notin f\left(\mathcal{E} \bigcup \mathbf{S}(w)\right) \text{ for each } 0 < w < \frac{\pi}{\gamma}.$$

An application of MATLAB subroutines shows that, for example, the family (6.6.1) is stable for $\gamma = 0.01$, and unstable for $\gamma = 0.05$. Images of principal points at the critical frequencies for the above values of γ are given next.

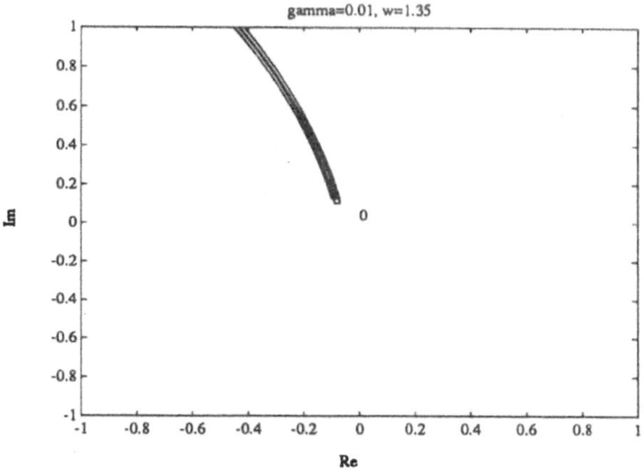

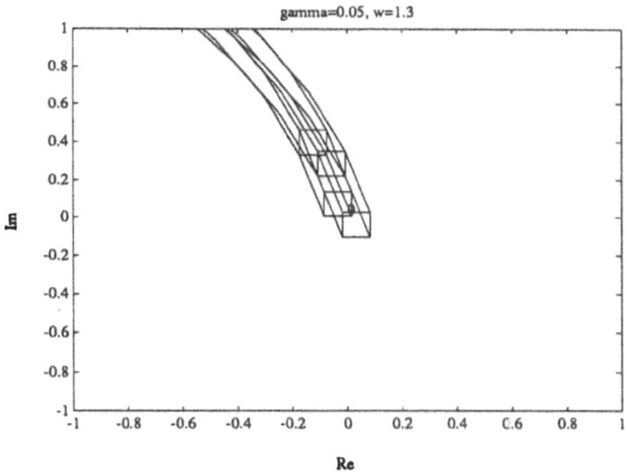

6.7. Eight Edges Theorem

This section illustrates a possible application of a quasipolynomial "extreme point" result of Kharitonov and Zhabko [KZ]. For a detailed description of the results we refer the reader to [HKKZ].

Motivated by the result of Kharitonov and Zhabko we focus on quasipolynomials with complex coefficients and **positive** shifts τ_i

$$\mathfrak{q}(s) = \sum_{n,k=0}^{N,K} t_{n,k} s^n e^{\tau_k s}, \quad 0 = \tau_0 < \tau_1 < \ldots < \tau_K.$$

Theorem 6.7.1 (Kharitonov and Zhabko.) Let

$$\mathbf{q}^0(s) = \sum_{n,k=0}^{N,K} t^0_{n,k} s^n e^{\tau_k s}, \text{ and } \mathbf{q}^1(s) = \sum_{n,k=0}^{N,K} t^1_{n,k} s^n e^{\tau_k s}, \quad t^i_{N,K} \neq 0, \ i = 0, 1$$

be two quasipolynomials with complex coefficients. If for each $w \in \mathbf{R}$

$$\frac{d}{dw}\arg\left[\mathbf{q}^0(jw) - \mathbf{q}^1(jw)\right] \leq \frac{\tau_0 + \tau_K}{2}, \tag{6.7.1}$$

then stability of $\mathbf{q}^0(s)$ and $\mathbf{q}^1(s)$ implies stability of the quasipolynomial segment

$$(1 - \lambda)\,\mathbf{q}^0(s) + \lambda \mathbf{q}^1(s).$$

In what follows we apply Theorem 6.7.1 to a complex coefficient weighted diamond quasipolynomial family described below.

$$\mathbf{Q} = \left\{ \mathbf{q}(s, \mathbf{x}, \mathbf{y}, \boldsymbol{\tau}) = \sum_{n,k=0}^{N,K} \left(\left[\underline{a}_{n,k} + \alpha_{n,k} x_{n,k}\right] + j \left[\underline{b}_{n,k} + \beta_{n,k} y_{n,k}\right] \right) s^n e^{\tau_k s}, \right\}, \tag{6.7.2}$$

where

$$(\mathbf{x}, \mathbf{y}) \in \mathbf{D}, \ \mathbf{D} = \left\{ (\mathbf{x}, \mathbf{y}) \ : \ \sum_{n,k} |x_{n,k}| + \sum_{n,k} |y_{n,k}| \leq 1 \right\}, \tag{6.7.3}$$

$$\tau_0 = 0, \quad 0 < \underline{\tau_k} \leq \tau_k \leq \overline{\tau_k}, \ k = 1, \ldots, K, \quad \overline{\tau_k} < \underline{\tau_K}, \ k = 1, \ldots, K-1,$$

and

$$\begin{aligned} &\alpha_{0,0} = \alpha_{N,0} = \beta_{0,0} = \beta_{N,0} = r\gamma, \quad r \geq \sqrt{2}, \\ &\gamma = \max\{\alpha_{n,k}, \ \beta_{n,k}\} \qquad\qquad (n, k) \neq (0, 0), (N, 0). \end{aligned} \tag{6.7.4}$$

Conditions (6.7.4) imply that for fixed $(w, \boldsymbol{\tau})$ the value set

$$\mathcal{Q}_{w,\boldsymbol{\tau}} = \{\mathbf{q}(jw, \mathbf{x}, \mathbf{y}, \boldsymbol{\tau}) \ : \ (\mathbf{x}, \mathbf{y}) \in \mathbf{D}\}$$

is a "rotated" square whose boundary is made up of the four segments

$$\begin{aligned} &\lambda\left[\mathbf{q}(jw, \mathbf{0}, \mathbf{0}, \boldsymbol{\tau}) \pm r\gamma\right] + (1 - \lambda)\left[\mathbf{q}(jw, \mathbf{0}, \mathbf{0}, \boldsymbol{\tau}) \pm jr\gamma\right], && \text{when } |w| \leq 1, \\ &\lambda\left[\mathbf{q}(jw, \mathbf{0}, \mathbf{0}, \boldsymbol{\tau}) \pm r\gamma w^N\right] + (1 - \lambda)\left[\mathbf{q}(jw, \mathbf{0}, \mathbf{0}, \boldsymbol{\tau}) \pm jr\gamma w^N\right], && \text{when } 1 \leq |w|. \end{aligned}$$

A straightforward computation shows that for fixed $\boldsymbol{\tau}$ the quasipolynomials

$$\mathbf{q}^0(s) = \mathbf{q}(s, \mathbf{0}, \mathbf{0}, \boldsymbol{\tau}) \pm r\gamma s^n \text{ and } \mathbf{q}^1(s) = \mathbf{q}(s, \mathbf{0}, \mathbf{0}, \boldsymbol{\tau}) \pm jr\gamma s^n$$

satisfy condition (6.7.1). Hence stability of the eight quasipolynomial families with constant coefficients and interval delays

$$\begin{aligned} &\mathbf{q}(s, \mathbf{0}, \mathbf{0}, \boldsymbol{\tau}) \pm r\gamma, \quad && \mathbf{q}(s, \mathbf{0}, \mathbf{0}, \boldsymbol{\tau}) \pm jr\gamma, \\ &\mathbf{q}(s, \mathbf{0}, \mathbf{0}, \boldsymbol{\tau}) \pm r\gamma s^N, \quad && \mathbf{q}(s, \mathbf{0}, \mathbf{0}, \boldsymbol{\tau}) \pm jr\gamma s^N, \end{aligned} \quad 0 \leq \underline{\tau}_k \leq \tau_k \leq \overline{\tau}_k \tag{6.7.5}$$

guarantees stability of the edges. Assumption 6.1.1 that yields the "zero exclusion" criterion can be simplified as follows:

Assumption 6.7.1 (V.L. Kharitonov) For each $(\mathbf{x},\mathbf{y}) \in \mathbf{D}$ one has

$$\left| \left[\underline{a}_{N,K} + \alpha_{N,K}x_{N,K}\right] + j\left[\underline{b}_{N,K} + \beta_{N,K}y_{N,K}\right] \right| >$$

$$\sum_{k=0}^{K-1} \left| \left[\underline{a}_{N,k} + \alpha_{N,k}x_{N,k}\right] + j\left[\underline{b}_{N,k} + \beta_{N,k}y_{N,k}\right] \right|.$$

The value set

$$Q_{0,\tau} = \left\{ \sum_{k=0}^{K} \left[\underline{a}_{0,k} + \alpha_{0,k}x_{0,k}\right] + j\left[\underline{b}_{0,k} + \beta_{0,k}y_{0,k}\right],\ (\mathbf{x},\mathbf{y}) \in \mathbf{D} \right\}$$

is a rotated square that does not depend on τ. Necessary and sufficient stability conditions are given next.

Theorem 6.7.2 Suppose that Assumption 6.7.1 holds. The quasipolynomial family (6.7.2) is stable if and only if

1. $\left|\sum_{k=0}^{K} \underline{a}_{0,k}\right| + \left|\sum_{k=0}^{K} \underline{b}_{0,k}\right| > r\gamma.$
2. The *eight* quasipolynomial families (6.7.5) with *constant* coefficient are stable.

For quasipolynomial families (6.7.2) with commensurate delays, i.e., $\tau_k = k\tau$, and $\underline{\tau} \leq \tau \leq \overline{\tau}$ the corresponding modification of Assumption 6.7.1 is, for example, the following.

Assumption 6.7.2 (V.L. Kharitonov) The polynomial family

$$\mathbf{P} = \left\{ \sum_{k=0}^{K} \left(\left[\underline{a}_{N,k} + \alpha_{N,k}x_{N,k}\right] + j\left[\underline{b}_{N,k} + \beta_{N,k}y_{N,k}\right] \right) z^k \ :\ (\mathbf{x},\mathbf{y}) \in \mathbf{D} \right\}$$

is Schur stable. That is for each $\mathbf{p}(z) \in \mathbf{P}$ one has $\mathbf{p}(z) \neq 0$ when $|z| \geq 1$.

Robust stability conditions for the quasipolynomial family with commensurate delays are given below.

Theorem 6.7.3 Suppose that Assumption 6.7.2 holds. The quasipolynomial family (6.7.2) with commensurate delays is stable if and only if

1. $\left|\sum_{k=0}^{K} \underline{a}_{0,k}\right| + \left|\sum_{k=0}^{K} \underline{b}_{0,k}\right| > r\gamma.$
2. The *eight one* parameter quasipolynomial families with *constant* coefficients

$$\begin{matrix} \mathbf{q}(s,0,0,\tau) \pm r\gamma, & \mathbf{q}(s,0,0,\tau) \pm jr\gamma, \\ \mathbf{q}(s,0,0,\tau) \pm r\gamma s^N, & \mathbf{q}(s,0,0,\tau) \pm jr\gamma s^N, \end{matrix} \quad 0 \leq \underline{\tau} \leq \tau \leq \overline{\tau}.$$

Necessary and sufficient stability conditions for quasipolynomial families with constant coefficients and interval delays have been recently reported by Boese [Boe2]. An application of the conditions to the eight families described above leads to an additional computationally tractable numerical method for checking stability of the family (6.7.2).

Necessary and sufficient robust stability conditions for quasipolynomial families with real interval coefficients and interval commensurate delays have been recently derived and reported by Kogan and Leizarowitz [KL2]. The procedure introduced in [KL2] requires checking positivity of a finite number of functions of w only.

We conclude with an example of an extreme point result borrowed from [HKKZ] .

Example 6.7.1 Consider the diamond family $\mathbf{Q}$ with two *constant* delays:

$$\left\{ \mathbf{q}(s,\mathbf{x},\mathbf{y},\boldsymbol{\tau}) = \begin{array}{l} [(1+\alpha_{0,0}x_{0,0}+\boldsymbol{j}\beta_{0,0}y_{0,0})\,e^{\tau_0 s} + (2+\alpha_{0,1}x_{0,1}+\boldsymbol{j}\beta_{0,1}y_{0,1})\,e^{\tau_1 s}] + \\ [(\alpha_{1,0}x_{1,0}+\boldsymbol{j}\beta_{1,0}y_{1,0})\,e^{\tau_0 s} + (1+\alpha_{1,1}x_{1,1}+\boldsymbol{j}\beta_{1,1}y_{1,1})\,e^{\tau_1 s}]\,s \end{array} \right\},$$

where

$$\alpha_{0,0} = \beta_{0,0} = \alpha_{1,0} = \beta_{1,0} = \frac{1}{2}, \text{ and } \alpha_{0,1} = \beta_{0,1} = \alpha_{1,1} = \beta_{1,1} = \frac{1}{4},$$

and the delays are $\tau_0 = 0$ and $\tau_1 = 1$. The weights satisfy (6.7.4), and since

$$\min_{(\mathbf{x},\mathbf{y})\in\mathbf{D}} |1+\alpha_{1,1}x_{1,1}+\boldsymbol{j}\beta_{1,1}y_{1,1}| \geq \frac{3}{4} > \frac{1}{2} \geq \max_{(\mathbf{x},\mathbf{y})\in\mathbf{D}} |\alpha_{1,0}x_{1,0}+\boldsymbol{j}\beta_{1,0}y_{1,0}|$$

there is no degree dropping in the right half plane (see Assumption 6.7.1). The eight quasipolynomials that determine stability of the entire family are

$$\begin{array}{ll} 1+(2+s)\,e^s \pm \tfrac{1}{2}, & 1+(2+s)\,e^s \pm \boldsymbol{j}\tfrac{1}{2}, \\ 1+(2+s)\,e^s \pm \tfrac{1}{2}s, & 1+(2+s)\,e^s \pm \boldsymbol{j}\tfrac{1}{2}s. \end{array}$$

When $s = \sigma + \boldsymbol{j}w$ and $\sigma \geq 0$ one has

$$|(2+s)\,e^s| = |(2+\sigma+\boldsymbol{j}w)\,e^\sigma| \geq 2+\sigma \geq 2.$$

This shows that the first six quasipolynomials and the nominal quasipolynomial

$$\mathbf{q}(s,0,0,\boldsymbol{\tau}) = 1+(2+s)\,e^s$$

are stable. To show that the quasipolynomial $1+(2+s)\,e^s+\boldsymbol{j}\frac{1}{2}s$ is stable, we apply the "zero exclusion" criterion to the quasipolynomial segment

$$\lambda\,[1+(2+s)\,e^s] + (1-\lambda)\left[1+(2+s)\,e^s+\boldsymbol{j}\frac{1}{2}s\right], \qquad 0 \leq \lambda \leq 1$$

and show that for each real w one has

$$0 \notin \lambda \left[1 + (2 + jw)\, e^{jw}\right] + (1 - \lambda) \left[1 + (2 + jw)\, e^{jw} - \frac{w}{2}\right], \qquad 0 \leq \lambda \leq 1. \quad (6.7.6)$$

If condition (6.7.6) is violated, the complex numbers

$$1 + (2 + jw)\, e^{jw}, \text{ and } 1 + (2 + jw)\, e^{jw} - \frac{w}{2} \qquad (6.7.7)$$

must be proportional. The complex numbers are proportional only when $(2 + jw)\, e^{jw}$ is real, and $(2 + jw)\, e^{jw} = \pm\sqrt{4 + w^2}$. A straightforward computation involving quadratic inequalities shows that when the complex numbers (6.7.7) are proportional they are real numbers of the same sign. This shows that (6.7.6) holds, and the quasipolynomial $1 + (2 + s)\, e^s + j\frac{1}{2}s$ is stable. A similar argument shows that the quasipolynomial $1 + (2 + s)\, e^s - j\frac{1}{2}s$ is stable. Stability of the entire quasipolynomial family follows now from Theorem 6.7.2.

Chapter 7

Convexity of Frequency Response Arcs Associated with Hurwitz Quasipolynomials

Geometry of the frequency response of stable rational functions has been recently under scrutiny by a number of authors. Horowitz and Ben–Adam [HBA], Bartlett [Bart], and Tesi, Vicino and Zappa [TVZ] investigated the *clockwise* properties of rational functions. Apart from its significance in analyzing frequency plots, the clockwise property is of interest also in the area of absolute stability of nonlinear systems. In particular, the results given in [TVZ] lead to a new class of systems for which the well known Kalman conjecture is true (see e.g., [Š]).

A different line of research concerning *convexity* properties of a frequency response arc associated with a Hurwitz polynomial has been initiated by Hamann and Barmish [HaB]. Roughly speaking the results of Hamann and Barmish can be described as follows. A frequency response arc of a polynomial $\mathbf{p}(s)$ is a plot

$$\{\psi(w) = \mathbf{p}(jw) \; : \; -\infty < w < \infty\}.$$

A part of the plot $\{\psi(w) \; : \; w_- \leq w \leq w_+\}$ is a *proper* frequency response arc if the net phase change of $\psi(w)$ does not exceed 180 degrees. A proper frequency response arc is convex if for any two distinct frequencies w_1, $w_2 \in [w_-, w_+]$ the arc does not intersect the interior of the triangle with the vertices 0, $\psi(w_1)$, and $\psi(w_2)$.

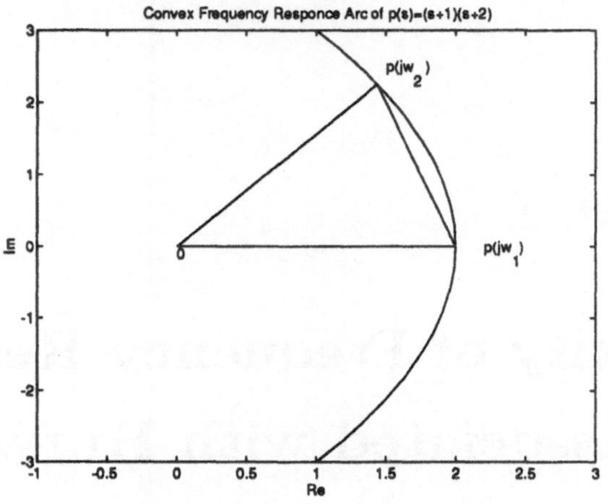

The main result in [HaB] shows that proper frequency response arcs associated with a Hurwitz polynomial are convex. It is also shown in [HaB] that the same result holds for any Ω stable polynomial, where Ω is a *convex* region in the complex plane. Furthermore, Hamann and Barmish show that the convexity of proper frequency response arcs leads to extreme point results for a certain class of linear systems.

In this Chapter following Cohen and Kogan [CK1] we set up a general formalism for treating argument increase and convexity in terms of angular derivatives, in analogy of the first and second derivative test for checking increase and convexity for real valued functions. Using this formalism, we obtain a short and elegant proof of argument increase and convexity for Hurwitz polynomials. We then proceed to prove the same two properties for a large class of entire functions with no zeros in the closed right half plane, namely Hurwitz stable functions of exponential type. This class contains Hurwitz quasipolynomials as a subset.

7.1. Convexity for Entire Functions

We start with a few formal definitions of frequency response arcs, their argument increase and convexity properties. These definitions are based on the original definitions introduced by Hamann and Barmish [HaB].

Definition 7.1.1 *Let* $[w_-, w_+] \subseteq [-\infty, +\infty]$ *be given. A generalized frequency response arc is a continuous mapping* $\psi : [w_-, w_+] \to \mathbf{C} \setminus \{0\}$.

Obviously, every transfer function $\mathbf{f}(s)$ without poles or zeros on the imaginary interval $[jw_-, jw_+]$ defines a frequency response arc via $\psi(w) = \mathbf{f}(jw)$.

Definition 7.1.2 *A generalized frequency response arc ψ is proper if the arc $\{\psi(w) : w_- \le w \le w_+\}$ is contained in an open half plane of the complex plane, not containing the origin.*

Definition 7.1.3 *A generalized frequency response arc ψ is argument increasing if $\arg \psi(w)$ is an increasing function of w.*

Definition 7.1.4 *A proper generalized frequency response arc ψ is pseudo–convex if the arc does not intersect the interior of the triangle with vertices 0, $\psi(w_-)$, and $\psi(w_+)$.*

Definition 7.1.5 *A generalized frequency response arc ψ is convex if every proper sub-arc of ψ is pseudo–convex.*

The difference between the last two definitions is illustrated below:

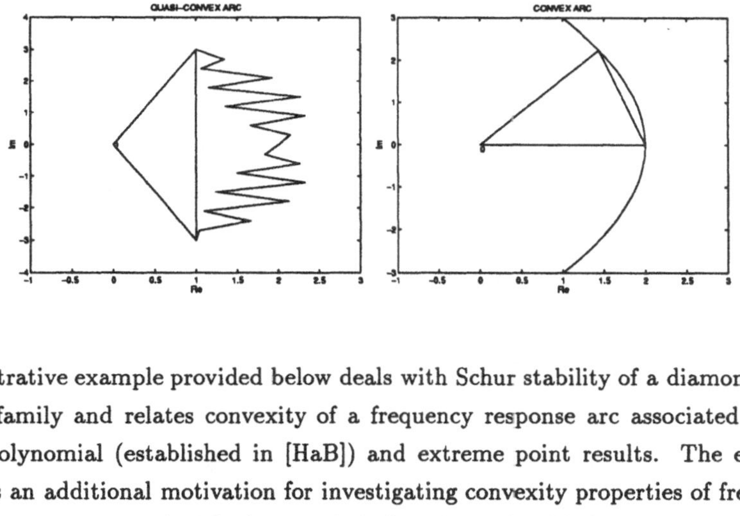

An illustrative example provided below deals with Schur stability of a diamond polynomial family and relates convexity of a frequency response arc associated with a Schur polynomial (established in [HaB]) and extreme point results. The example provides an additional motivation for investigating convexity properties of frequency response arcs associated with characteristic functions of stable linear systems.

Example 7.1.1 Consider the diamond polynomial family

$$\mathbf{P}_\gamma = \left\{ \sum_{k=0}^{5} [a_k + j b_k] z^k \right\}$$

where the coefficients belong to the diamond

$$\sum_{k=0}^{5} \left| \frac{a_k - a_k^0}{\alpha_k} \right| + \sum_{k=0}^{5} \left| \frac{b_k - b_k^0}{\beta_k} \right| \le \gamma,$$

the nominal polynomial is

$$\mathbf{p}^0(z) = \sum_{k=0}^{5}[a_k^0 + jb_k^0]z^k = 0.45z - (6 + j0.6)z^2 + (18 + j8)z^3 - (40 + j20)z^4 + 100z^5,$$

and the weights are

$$\alpha_0 = \beta_0 = 1,\ \alpha_k = \beta_k = \frac{1}{\sqrt{2}},\ k = 1, \ldots, 5.$$

The nominal polynomial is Schur stable, hence the corresponding frequency response arc is convex. It is easy to see that the value set of the polynomial family is a rotated square with vertices at the points $\mathbf{p}^0\left(e^{j\theta}\right) \pm \gamma$ and $\mathbf{p}^0\left(e^{j\theta}\right) \pm j\gamma$, $0 \le \theta \le 2\pi$. The polynomial family $\mathbf{P}_\gamma$ is stable if and only if the origin does not belong to the value sets. Equivalently, the polynomial family $\mathbf{P}_\gamma$ is stable if and only if the rotated square with vertices at the points $\pm\gamma$ and $\pm j\gamma$ does not intersect the frequency response of the nominal polynomial. Hence as γ increases the first contact point between the rotated square and the frequency response occurs at one of the extreme points. The simple graphical test immediately yields that the critical value is $\gamma^* = 63$.

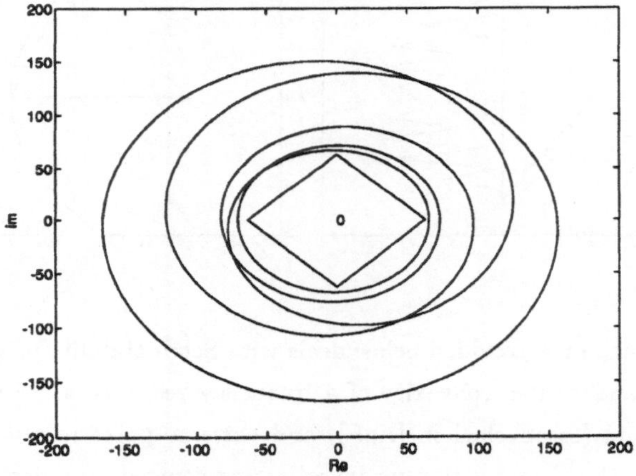

The graphical test reminds the Tsypkin–Polyak locus (see p. 54). In general, however, the frequency response of Tsypkin–Polyak hodograph is nonconvex, and it does not generate extreme point results. On the other hand extreme point results reported by Rantzer [Ra1] require convexity of both the stability domain and its reciprocal, a condition that obviously fails in the Schur stability case. Furthermore, while Rantzer's result is valid for boxes of arbitrary sides, diamond extreme point results are extremely sensitive to the choice of weights (see Section 2.8).

It is natural to ask whether argument increase and arc convexity can be obtained in a more general context. It turns out that large classes of entire functions whose zeros are all in the open left half plane enjoy the same two properties. Classes of this type were discussed at length in the classical book [Lev] by B.Ja. Levin, where the stability region was taken to be the upper (rather than the left) half plane. Our definitions will conform with his, except for the trivial adaptation of the region of stability, i.e. a simple 90 degree rotation in the s domain.

Definition 7.1.6 *We shall call an entire function* $\mathbf{f}(s)$ *Hurwitz (weakly Hurwitz) if its roots are in the open (respectively closed) left half plane.*

Definition 7.1.7 *We shall say that* $\mathbf{f}$ *is in the class* HB $(\overline{HB})$ *if it is Hurwitz (respectively weakly Hurwitz) and in addition* $\mathrm{Re}\ s > 0$ *implies* $|\mathbf{f}(s)| \geq |\mathbf{f}(-\bar{s})|$. *The class* HB *is known as the Hermite–Biehler class.*

Definition 7.1.8 *The class* P *consists of all* $\overline{HB}$ *functions* $\mathbf{f}(s)$ *which are of exponential type, i.e.* $|\mathbf{f}(s)|$ *is bounded by* $Me^{\sigma|s|}$ *for some* $M, \sigma \in [0, \infty)$.

Definition 7.1.9 *The class* P^* *consists of all* $\overline{HB}$ *functions of the special form* $\mathbf{f}(s) = e^{\gamma s^2} g(s)$, *where* g *is a function of genus not larger than* 1, *and* $\gamma \geq 0$ *(for a precise definition of genus, see* [Lev]).

Stability theory for entire functions hinges crucially on these definitions. In the case of rational functions, all these definitions collapse to the sets of Hurwitz or weakly Hurwitz polynomials. For general entire functions, however, they are all different: the class of weakly Hurwitz polynomials, P, P^*, $\overline{HB}$, and the class of weakly Hurwitz entire functions represent a chain of strict inclusions.

The tremendous importance of the class P^* in stability theory is expressed by the following result ([Lev], p. 334):

Theorem 7.1.1 P^* consists of all the entire functions which are limits of weakly Hurwitz polynomials.

Limit is understood here in the sense of uniform convergence on every bounded set in **C**. The weakly Hurwitz stability assumption on the approximating polynomials is inessential: if $\mathbf{f} \in P^*$ is the limit of the weakly Hurwitz polynomials $\mathbf{p}_n(s)$ then it is also the limit of the strongly Hurwitz polynomials $\mathbf{p}_n^*(s) := \mathbf{p}_n(s + 1/n)$.

Theorem 7.1.1 implies that basically all the interesting properties of weakly Hurwitz polynomials are shared by all functions of class P^*. For example, this class is closed under differentiation; and the interlacing characterization in terms of the real and imaginary parts is still available ([Lev], p. 335). In fact, it is shown in [Lev] that interlacing extends to the larger class $\overline{HB}$.

We are interested in yet another corollary of Theorem 7.1.1:

Theorem 7.1.2 The frequency response arc of every function in P^* is argument non-decreasing and convex.

In fact, in Section 7.3 it will be shown that more can be said about the behavior of these arcs. To prove Theorem 7.1.2 we show that the following result holds:

Lemma 7.1.1 *Let* $\psi_n : \mathbf{R} \to \mathbf{C}$, $n = 1, 2, \ldots$ *be a sequence of convex generalized frequency response arcs. If there exists a function* $\psi(w)$ *such that for each real* w *one has* $\psi(w) \neq 0$, *and* $\lim_{n\to\infty} \psi_n(w) = \psi(w)$, *then* $\psi(w)$ *is a convex generalized frequency response arc.*

The proof will be given in Section 7.3.

It is quite possible that argument increase and convexity can also be established for classes larger than P^*. However, such results cannot be deduced from classical Hurwitz polynomial theory by means of direct approximation.

As an important application of Theorem 7.1.2, consider the set of Hurwitz quasipolynomials:

Definition 7.1.10 *A quasipolynomial*

$$\mathsf{q}(s) = \sum_{n,k=0}^{N,K} a_{n,k} s^n e^{\tau_k s}, \qquad 0 \leq \tau_0 < \ldots < \tau_K$$

is Hurwitz if $\mathsf{q}(s) \neq 0$ *when* $\operatorname{Re} s \geq 0$.

Corollary 7.1.1 *The frequency response arc of a Hurwitz quasipolynomial is argument nondecreasing and convex.*

Proof: It is known that P^* contains P. Therefore, due to Theorem 7.1.2 it is enough to show that Hurwitz quasipolynomials are in the set P. Indeed, they are Hurwitz, and of exponential type. In order to show that they belong to HB, we have to show that

$$\operatorname{Re} s > 0 \text{ implies } |\mathsf{q}(s)| \geq |\mathsf{q}(-\bar{s})|. \tag{7.1.1}$$

It has been shown by Schwengeler [Sc] that a Hurwitz quasipolynomial can be represented by the simple Hadamard product (see [Ru])

$$\mathbf{q}(s) = \mathbf{q}(0)e^{(\tau_0+\tau_K)s/2}\prod_{i=1}^{\infty}\left(1-\frac{s}{s_i}\right). \tag{7.1.2}$$

where s_i are the roots. The convergence properties of infinite products of this type are well understood (for a detailed discussion see also [KZ] and [Lev].) Since Re $s_i < 0$, each factor in this product has the property (7.1.1). Therefore, the whole product has this property, and so $\mathbf{q} \in P$, completing the proof.

In fact, a similar argument shows that every Hurwitz entire function of exponential type is in the class HB, see [Lev].

It is tempting to assert that the product of two convex arcs $\psi(w) = \phi_1(w)\phi_2(w)$ is convex. Since each

$$\psi_n(w) = \mathbf{q}(0)e^{\tau \boldsymbol{j} w}\prod_{i=1}^{n}\left(1-\frac{\boldsymbol{j}w}{s_i}\right)$$

is a product of two convex proper generalized frequency response arcs $\phi_1(w) = e^{\tau \boldsymbol{j} w}$, and $\phi_2(w) = \mathbf{q}(0)\prod_{i=1}^{n}\left(1-\frac{\boldsymbol{j}w}{s_i}\right)$ this result would immediately lead to convexity of $\psi_n(w)$. However, as we show by an example, the assertion is in general false.

Example 7.1.2 Let $\phi_1(w) = 1 + w + (100 + \boldsymbol{j})w^2$, and $\phi_2(w) = 1 + \boldsymbol{j}w$. When $0.01 \leq w \leq 0.05$ the functions ϕ_1 and ϕ_2 define convex proper generalized frequency response arcs. On the other hand the product $\{\phi_1(w)\phi_2(w) \ : \ 0.01 \leq w \leq 0.05\}$ is *nonconvex.*

The plot of $\phi_2(w)$ is a vertical line, the plots of $\phi_1(w)$ and $\phi_1(w)\phi_2(w)$ for $0.01 \leq w \leq 0.05$ are given below.

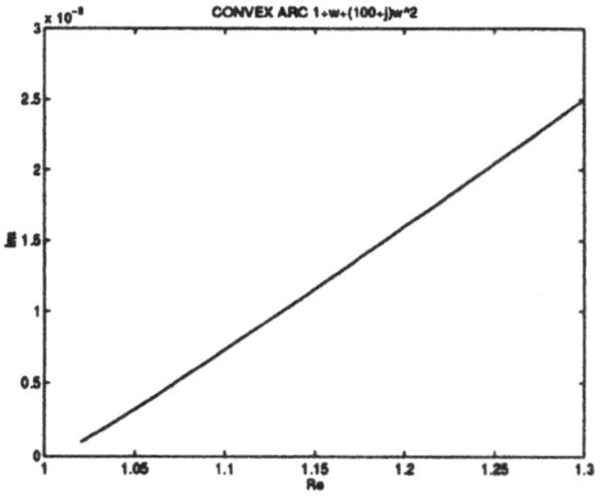

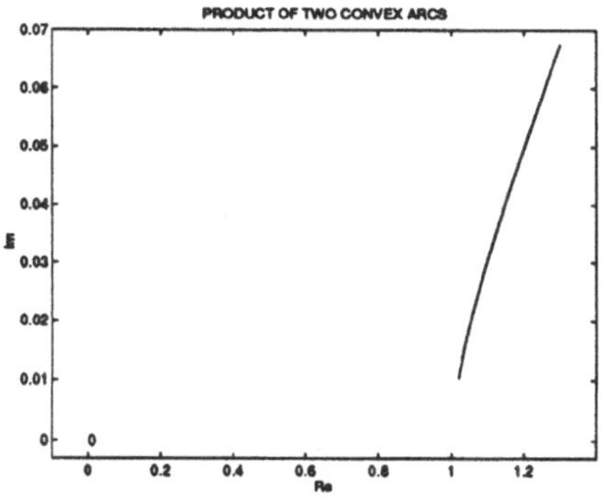

7.2. Angular Derivative Formalism

In this section we reformulate argument increase and arc convexity in terms of certain inner products, which we interpret as first and second angular derivatives. The resulting formulation resembles the first and second derivative test for increase and convexity of *real valued* functions.

For sufficiently smooth real valued functions the sign of various derivatives has *local* meaning. Positive first or second derivative at some point implies that the function in question is increasing, or convex, in some neighborhood of the given point. Higher derivatives may also be interpreted locally, in terms of more subtle characteristics. Moreover, when the kth derivative vanishes, the relevant local behavior can only be determined by examining higher order derivatives.

The same interpretation is available for frequency response arcs in the complex plane, under the appropriate smoothness assumptions (in all foreseeable applications, the arcs in question will be analytic, hence as smooth as needed). This interpretation is explained below.

Definition 7.2.1 *The k-th angular derivative of the generalized frequency response arc $\psi(w)$ is defined as*

$$\langle j\psi^{(k-1)}(w), \psi^{(k)}(w)\rangle.$$

(For the definition of the dot product of two complex numbers see p. 29, Definition 2.8.1.) The "first derivative test"

$$\langle j\psi(w), \psi'(w)\rangle > 0 \tag{7.2.1}$$

guarantees local argument increase. In geometric terms, this condition implies that $\operatorname{Im} \dfrac{\psi'(w)}{\psi(w)} > 0$, i.e., the tangent vector $\psi'(w)$ is located to the "left" of the location vector $\psi(w)$. As in the analogous case of real valued functions, argument increase may occur even if the first angular derivative vanishes, depending on the signs of higher angular derivatives.

Next, the "second derivative test" is given by

$$\langle j\psi(w), \psi'(w)\rangle \cdot \langle j\psi'(w), \psi''(w)\rangle > 0 \tag{7.2.2}$$

This condition expresses local convexity of ψ at w *regardless of argument orientation.* Assuming local argument increase (i.e. (7.2.1)), this condition simplifies to

$$\langle j\psi'(w), \psi''(w)\rangle > 0 \tag{7.2.3}$$

Again, arc convexity is still possible if the second angular derivative vanishes.

Of course, the local definitions given in this section are in complete agreement with the global definitions given in Section 2. Namely,

Lemma 7.2.1 *Let $\{\psi(w) : w_- \leq w \leq w_+\}$ be a generalized frequency response arc. If (7.2.1) holds for each $w \in [w_-, w_+]$, then ψ is argument increasing. Similarly, if, in addition, (7.2.2) holds for each $w \in [w_-, w_+]$, then ψ is arc convex.*

Proof: The argument increase is clear. To prove the convexity we have to show that the arc does not intersect the interior of the triangle with vertices 0, $\psi(w_-)$, and $\psi(w_+)$. This geometric condition is expressed by the inequality

$$\langle j\,[\psi(w_+) - \psi(w_-)]\,, \psi(w) - \psi(w_-)\rangle \leq 0 \text{ for each } w \in [w_-, w_+]. \tag{7.2.4}$$

To prove the lemma we assume that (7.2.4) is violated, i.e.,

$$\max_{w_- \leq w \leq w_+} \langle j\,[\psi(w_+) - \psi(w_-)]\,, \psi(w) - \psi(w_-)\rangle > 0. \tag{7.2.5}$$

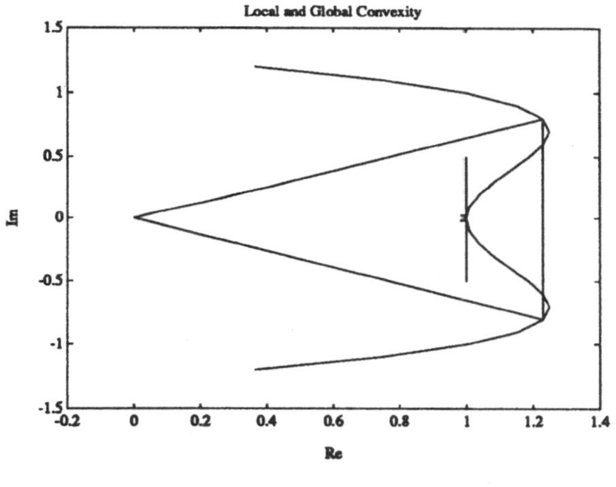

Let w_0 be a point where the maximum is attained, i.e.,

$$\begin{aligned}\langle j\,[\psi(w_+) - \psi(w_-)]\,, \psi(w_0) - \psi(w_-)\rangle &= \\ \max_{w_- \leq w \leq w_+} \langle j\,[\psi(w_+) - \psi(w_-)]\,, \psi(w) - \psi(w_-)\rangle &> 0.\end{aligned} \tag{7.2.6}$$

Then $\psi(w_+) - \psi(w_-) = \alpha\psi'(w_0)$, where α is a real constant. Furthermore, due to the increasing argument condition $\alpha > 0$. Let

$$c(w) = \langle \boldsymbol{j}\,[\psi(w_+) - \psi(w_-)]\,, \psi(w) - \psi(w_-)\rangle \tag{7.2.7}$$

Then

$$\begin{aligned} c(w_0) &= \alpha\langle \boldsymbol{j}\psi'(w_0), \psi(w_0) - \psi(w_-)\rangle, \\ c'(w_0) &= \alpha\langle \boldsymbol{j}\psi'(w_0), \psi'(w_0)\rangle = 0, \\ c''(w_0) &= \alpha\langle \boldsymbol{j}\psi'(w_0), \psi''(w_0)\rangle > 0. \end{aligned}$$

This shows that w_0 is a strict local minimum of $c(w)$. On the other hand due to (7.2.6)

$$\begin{aligned} c(w) &= \langle \boldsymbol{j}\,[\psi(w_+) - \psi(w_-)]\,, \psi(w) - \psi(w_-)\rangle \\ &\leq \langle \boldsymbol{j}\,[\psi(w_+) - \psi(w_-)]\,, \psi(w_0) - \psi(w_-)\rangle = c(w_0). \end{aligned}$$

This contradiction completes the proof.

The lemma provides a sufficient condition only for argument increase and arc convexity. A straight line $\psi(w) = 1 + \boldsymbol{j}w$ is a convex frequency response arc whose second angular derivative $\langle \boldsymbol{j}\psi'(w), \psi''(w)\rangle = 0$.

7.3. Computation of Angular Derivatives

Our analysis hinges on the concrete computation of the first angular derivative for an arbitrary polynomial $\mathbf{p}(s) = (s - s_1)\ldots(s - s_n)$. For each $s_0 \in \mathbf{C}$ which is not a zero of $\mathbf{p}$ one has

$$\langle \mathbf{p}(s_0), \mathbf{p}'(s_0)\rangle = \left\langle \mathbf{p}(s_0), \mathbf{p}(s_0)\sum_{i=1}^{n}\frac{1}{s_0 - s_i}\right\rangle =$$

$$|\mathbf{p}(s_0)|^2 \sum_{i=1}^{n}\langle 1, \frac{1}{s_0 - s_i}\rangle = |\mathbf{p}(s_0)|^2 \sum_{i=1}^{n} \mathrm{Re}\,\frac{1}{s_0 - s_i}$$

If, in addition, $\mathbf{p}(s)$ is Hurwitz and $\mathrm{Re}\, s_0 \geq 0$, then $\mathrm{Re}\,\frac{1}{s_0 - s_i} > 0$ for each $i = 1, \ldots, n$, and we obtain the following result.

Lemma 7.3.1 *If* $\mathbf{p}(s)$ *is a Hurwitz polynomial and* $\mathrm{Re}\, s_0 \geq 0$, *then*

$$\langle \mathbf{p}(s_0), \mathbf{p}'(s_0)\rangle = |\mathbf{p}(s_0)|^2 \sum_{i=1}^{n} \mathrm{Re}\,\frac{1}{s_0 - s_i} > 0. \tag{7.3.1}$$

Argument increase and convexity can easily be established for Hurwitz polynomials using angular derivatives.

Theorem 7.3.1 If $\mathbf{p}(s)$ is a Hurwitz polynomial, then the frequency response arc $\psi(w) = \mathbf{p}(\boldsymbol{j}w)$, $w \in \mathbf{R}$, is argument increasing and convex.

Proof: By setting $s_0 = \boldsymbol{j}w$ in Lemma 7.3.1 we obtain that $\mathbf{p}(s)$ is argument increasing, by the first angular derivative test. Moreover, by Gauss Theorem (often referred to as Lucas Lemma, see e.g., [PS], p. 108), $\mathbf{p}'(s)$ is also Hurwitz, hence by the second angular derivative test $\mathbf{p}(s)$ is also arc convex.

Remark 7.3.1 Convexity of a frequency response arc associated with Ω stable polynomial when Ω is a convex subset of the complex plane is established in [HaB]. The second derivative technique immediately recovers this result (see [Ko2]).

In order to establish the more general Theorem 7.1.1, we have to consider the limit behavior of a sequence of frequency response arcs. Consider a proper generalized frequency response arc

$$\{\psi(w) = \mathbf{q}(\boldsymbol{j}w) \ : \ w_1 \leq w \leq w_2\},$$

and a sequence of proper generalized frequency response arcs

$$\{\psi_n(w) = \mathbf{q}_n(\boldsymbol{j}w) \ : \ w_1 \leq w \leq w_2\}.$$

For each $w \in [w_1, w_2]$ one has $\lim_{n\to\infty} \psi_n(w) = \psi(w)$, and

$$0 \geq \lim_{n\to\infty} \langle \boldsymbol{j}\,[\psi_n(w_2) - \psi_n(w_1)]\,, \psi_n(w) - \psi_n(w_1)\rangle = \langle \boldsymbol{j}\,[\psi(w_2) - \psi(w_1)]\,, \psi(w) - \psi(w_1)\rangle.$$

This shows convexity of the proper generalized frequency response arc $\psi(w)$, and completes the proof of Theorem 7.1.1.

Remark 7.3.2 An elementary proof of Theorem 7.1.1 based on convexity of the frequency response arcs

$$\psi_n(w) = \mathbf{q}(0)e^{\tau \boldsymbol{j}w} \prod_{i=1}^{n} \left(1 - \frac{\boldsymbol{j}w}{s_i}\right)$$

is given in [CK2]. The proof hinges on the following two interesting observations:

Lemma 7.3.2 *Let* $\mathbf{p}(s)$ *be a Hurwitz polynomial. For each* $\tau \geq 0$ *the polynomial* $\tau\mathbf{p}(s) + \mathbf{p}'(s)$ *is Hurwitz stable.*

Proof: Suppose that $\tau\mathbf{p}(s)+\mathbf{p}'(s)$ is not Hurwitz. Let s_0 be one of its unstable roots ($\mathrm{Re}\ s_0 \geq 0$). Then we get $\langle \mathbf{p}(s_0), \mathbf{p}'(s_0)\rangle = -\tau\,|\mathbf{p}(s_0)|^2 \leq 0$, contradicting Lemma 7.3.1.

Lemma 7.3.3 Let $\mathbf{p}(s)$ be a Hurwitz polynomial, and $\tau \geq 0$. If $\psi(w) = e^{\tau \boldsymbol{j} w}\mathbf{p}(\boldsymbol{j}w)$, then (7.2.2) holds.

Proof: A straightforward computation shows that the derivative of $\psi(w)$ is

$$\psi'(w) = \boldsymbol{j}e^{\tau \boldsymbol{j} w}\left[\tau\mathbf{p}(\boldsymbol{j}w) + \mathbf{p}'(\boldsymbol{j}w)\right],$$

and

$$\langle \boldsymbol{j}\psi(w), \psi'(w)\rangle = \tau|\mathbf{p}(\boldsymbol{j}w)|^2 + \langle \mathbf{p}(\boldsymbol{j}w), \mathbf{p}'(\boldsymbol{j}w)\rangle > 0.$$

We denote the polynomial $\tau\mathbf{p}(s)+\mathbf{p}'(s)$ by $\mathbf{p}_1(s)$. Due to Lemma 7.3.2 the polynomial $\mathbf{p}_1(s)$ is Hurwitz stable. Since

$$\psi'(w) = \boldsymbol{j}e^{\tau \boldsymbol{j} w}\mathbf{p}_1(\boldsymbol{j}w), \text{ and } \psi''(w) = -e^{\tau \boldsymbol{j} w}\left[\tau\mathbf{p}_1(\boldsymbol{j}w) + \mathbf{p}_1'(\boldsymbol{j}w)\right] \tag{7.3.2}$$

one has

$$\begin{aligned}
\langle \boldsymbol{j}\psi'(w), \psi''(w)\rangle &= \langle -e^{\tau \boldsymbol{j} w}\mathbf{p}_1(\boldsymbol{j}w), -e^{\tau \boldsymbol{j} w}\left[\tau\mathbf{p}_1(\boldsymbol{j}w) + \mathbf{p}_1'(\boldsymbol{j}w)\right]\rangle \\
&= \langle \mathbf{p}_1(\boldsymbol{j}w), \tau\mathbf{p}_1(\boldsymbol{j}w) + \mathbf{p}_1'(\boldsymbol{j}w)\rangle \\
&= \tau|\mathbf{p}_1(\boldsymbol{j}w)|^2 + \langle \mathbf{p}_1(\boldsymbol{j}w), \mathbf{p}_1'(\boldsymbol{j}w)\rangle > 0.
\end{aligned}$$

We have shown that the treatment of convex arcs in terms of angular derivatives, which we introduced, is analogous to the Hamann–Barmish treatment. It should be mentioned that a third, quite different, treatment also exists: the Minkowski functional, which is used in convexity theory (see e.g., [Ber]). The advantage of this approach is that convexity of a set can be characterized even when this set does not contain the origin.

In general, describing convexity of an arc in these terms may be quite tedious. However, in the special case of the complex plane, this can be done quite neatly. The interested reader is referred to [Lev] Section I.19.

7.4. Inner Frequency Response Set

The notion of inner frequency response set has been introduced in [HaB] in the context of Hurwitz polynomials. A variant of this definition, which is more suitable for work with non–rational functions, is given below:

Definition 7.4.1 *The geometric inner frequency response set of* ψ *consists of all points in the complex plane which can be connected to the origin via a linear segment that does not intersect* $\{\psi(w) \ : \ w_- \leq w \leq w_+\}$.

For comparison, we would also like to introduce a slightly different definition.

Definition 7.4.2 *The topological inner frequency response set of* ψ *consists of all points in the complex plane which can be connected to the origin via an arc that does not intersect* $\{\psi(w) \;:\; w_- \leq w \leq w_+\}$.

We shall at times abbreviate the term "inner frequency response set" to "inner set".

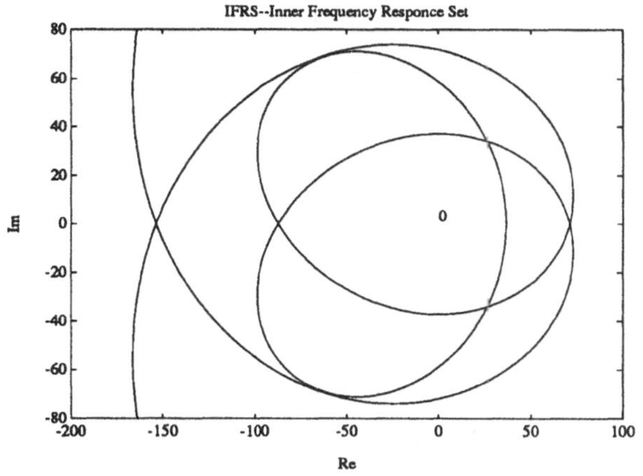

The geometric inner set is always star–shaped, the topological variant is merely connected, and may be a larger set. The following is a reformulation of the result of Hamann and Barmish [HaB] concerning inner sets.

Theorem 7.4.1 If $\mathbf{p}(s)$ is a Hurwitz polynomial the geometric inner frequency response set of $\psi(w)$ is a convex open subset of the complex plane that contains the origin in its interior.

Let us add that in this case the topological and geometric inner sets are identical; if the degree of $\mathbf{p}(s)$ is at least 3, the inner set is bounded.

It is interesting to examine to what extent these observations extend to other classes of stable entire functions. The geometric observations, i.e. the convexity and coincidence of the two inner sets, can be established in the set P^*, using approximation by Hurwitz polynomials. The topological observations, i.e. that the inner set is open, bounded, and contains the origin in its interior rather than on the boundary, are more delicate.

Let $\mathbf{f}(s)$ be any function $\mathbf{f}(s) \in P^*$ which is not a polynomial. Approximation by Hurwitz polynomials shows that the arc $\mathbf{f}(jw)$ winds counterclockwise around the origin an infinite number of times. Typically, it escapes to ∞ as w tends to $\pm\infty$. This, plus some work, shows that typically the inner set is bounded and its boundary is a finite union of proper sub–arcs. Thus, the inner set contains no point on this boundary, hence is open, and contains the origin in its interior.

Theorem 7.4.1 for polynomials follows easily from these considerations. Indeed, for a Hurwitz polynomial of degree at least 3, the arc escapes to ∞ and the boundary of the inner set is a union of a finite number of proper arcs.

However, for a relatively small subset of functions in P^*, the boundary of the inner set may be an infinite union of proper sub–arcs or isolated arc points; or even worse, it may contain points which are not on the arc at all, i.e. limit points. In this case, the inner set may fail to be open. One possible source for this behavior may be boundedness of the arc $\mathbf{f}(jw)$.

In such a case, the boundary of the inner set may contain the origin. This is a rare occurrence, which implies that $\mathbf{f}(s)$ is *barely stable*, i.e. the real parts of its zeros have 0 as an accumulation point.

In what follows we illustrate these difficulties in the context of Hurwitz quasipolynomials by examples.

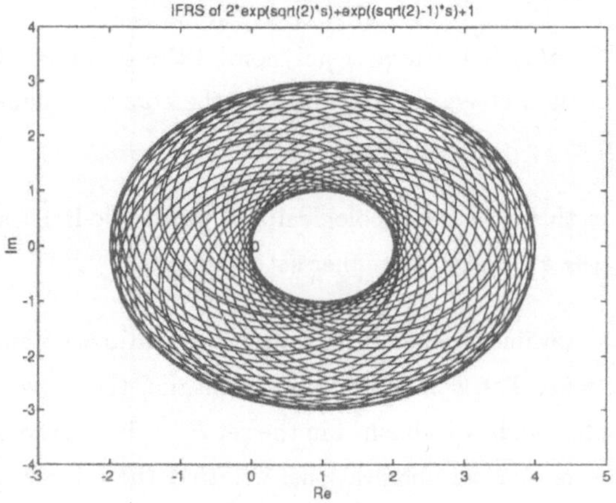

Example 7.4.1 Consider the quasipolynomial

$$\mathbf{q}(s) = 2e^{\sqrt{2}s} + e^{(\sqrt{2}-1)s} + 1 = e^{\sqrt{2}s}\left[2 + e^{-s} + e^{-\sqrt{2}s}\right].$$

The quasipolynomial $\mathbf{q}(s)$ is Hurwitz, and the frequency response arc $\mathbf{q}(jw)$ is convex. The inner set is convex but not open (in fact, it is closed). The origin is a cluster point of the arc (see p. 118, Example 6.1.1).

Next let us consider a general Hurwitz quasipolynomial

$$\mathbf{q}(s) = \left[\sum_{k=0}^{K} a_{0,k}e^{\tau_k s}\right] + \left[\sum_{k=0}^{K} a_{1,k}e^{\tau_k s}\right] s + \ldots + \left[\sum_{k=0}^{K} a_{N,k}e^{\tau_k s}\right] s^N$$

with "degree" N. First assume that $N = 0$.

If $N = 0$, the arc $\mathbf{q}(jw)$ is bounded (in fact, quasi–periodic, see e.g. [Lev]). If all the ratios $\tau_k/\tau_{k'}$ are rational (the so–called fully commensurate case), the arc is periodic, hence the origin is an interior point of the inner set. If τ_k are linearly independent over the rational number field, a necessary and sufficient condition for the origin to be in the interior is that the modulus of one of the coefficients $a_{0,k}$ is larger than the sum of all the other moduli. We shall call this condition "the coefficient dominance condition".

In Example 7.4.1 the delays are not fully commensurate and linearly dependent, the "weak coefficient dominance condition" holds, $\mathbf{q}(s)$ is barely stable, and the origin belongs to the boundary of the envelope spanned by the full arc.

Now consider a Hurwitz quasipolynomial $\mathbf{q}(s)$ with "degree" $N > 0$. Define its "leading coefficient" to be the quasipolynomial of "degree" 0, $\mathbf{q}_N(s) := \sum a_{N,k}e^{\tau_k s}$. If $\mathbf{q}_N(s)$ satisfies the coefficient dominance condition, it can be easily shown that for large w the arc $\mathbf{q}(jw)$ escapes to ∞, and the boundary of the inner set is a finite union of proper sub–arcs, hence the origin is an interior point of the inner set. A typical inner frequency response set of a simple Hurwitz quasipolynomial whose "leading coefficient" satisfies the coefficient dominance condition is shown below.

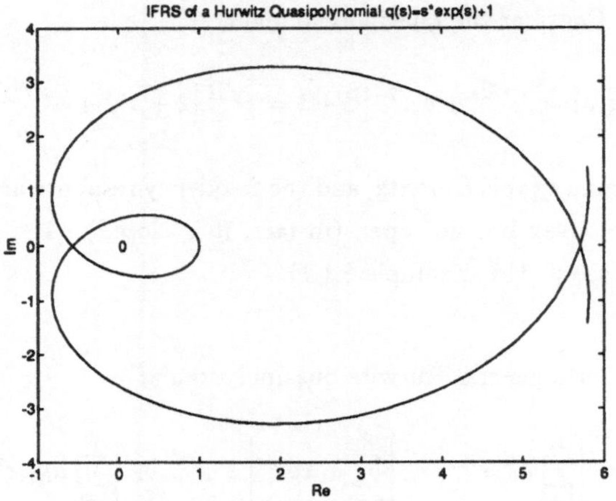

The coefficient dominance condition applied to the "leading coefficient" $\mathbf{q}_N(s)$ of $\mathbf{q}(s)$ prevents a drop in the "degree". It is typical in robustness analysis of quasipolynomial families (for detailed discussion see e.g., [HKZ2]).

Chapter 8

Epilogue

In this chapter we return briefly to the stability problem stated in Chapter 1. We remind the reader that a $n \times n$ matrix A is stable if and only if det $|sI - A| \neq 0$ when Re $s \geq 0$. We shall consider a polytope of matrices

$$\mathcal{A} = \left\{ A = \sum_{i=1}^{k} \alpha_i A_i \ : \ \sum_{i=1}^{k} \alpha_i = 1, \ \alpha_i \geq 0, \ i = 1, \ldots, k \right\}.$$

From a computational point of view stability verification of a polytope of matrices is known to be NP–hard (see [Nem]). Problems of this type are regarded in Computer Science as computationally intractable.

From a control theoretic point of view the robust synthesis problem with structured real parametric uncertainty is probably the most important and the most difficult one. This problem can be stated as follows: For two polytopes of matrices $\mathcal{A}$ and $\mathcal{B}$ find a matrix K such that the set of matrices

$$\{A + BK \ : \ A \in \mathcal{A}, \ B \in \mathcal{B}\}$$

is stable. As of today necessary and sufficient conditions for simultaneous stabilization of three and more plants is a non trivial problem (see e.g., [WF]). [1] For a comprehensive review of recent results concerning simultaneous stabilization we refer the reader to [Blo].

In the next two sections we show how applications of Lyapunov stability criterion lead to partial solutions for these two problems.

[1] A bottle of good French champagne is offered by the authors of [BG] to the first person to prove or disprove that *three* specific continuous systems are simultaneous stabilizable.

8.1. Lyapunov Criterion

We start with a standard material concerning Lyapunov stability theory.

Theorem 8.1.1 (Lyapunov). A matrix A is stable if and only if for any given positive–definite symmetric matrix Q there exists a positive–definite symmetric matrix P that satisfies

$$A'P + PA = -Q. \tag{8.1.1}$$

Proof: *Necessity:* Since A is stable $e^{A't}Qe^{At}$ is the sum of terms of the form $t^k e^{st}$ where Re $s < 0$. Hence the equation

$$P = \int_0^\infty e^{A't}Qe^{At}\, dt \tag{8.1.2}$$

defines a positive–definite and symmetric matrix P. By substitution

$$\begin{aligned} A'P + PA &= \int_0^\infty A'e^{A't}Qe^{At}\, dt + \int_0^\infty e^{A't}Qe^{At}A\, dt \\ &= \int_0^\infty \frac{d}{dt} e^{A't}Qe^{At}\, dt = e^{A't}Qe^{At}\, |_0^\infty = -Q. \end{aligned}$$

Sufficiency: Let $L(\mathbf{x}) = \mathbf{x}'P\mathbf{x}$. If $\mathbf{x}(t)$ is a solution of the equation

$$\dot{\mathbf{x}}(t) = A\mathbf{x}(t), \tag{8.1.3}$$

then

$$\frac{d}{dt}L(\mathbf{x}(t)) = \mathbf{x}'(t)\,[A'P + PA]\,\mathbf{x}(t) = -\mathbf{x}'(t)Q\mathbf{x}(t) < 0.$$

This shows that each solution $\mathbf{x}(t)$ of the system (8.1.3) converges to the origin as $t \to \infty$, and A is stable.

Corollary 8.1.1 *(Lyapunov, 1893). If A is stable, then for every matrix Q the Lyapunov equation (8.1.1) has a* unique *solution.*

Proof: For every Q, the integral (8.1.2) defines a solution of the n^2 linear equations (8.1.1) with respect to p_{ij}. Since for *every* choice of Q the system has a solution, this solution must be *unique.*

Corollary 8.1.2 *Suppose that A is stable. If P is a symmetric $n \times n$ matrix such that the matrix $A'P + PA$ is negative definite, then P is positive definite.*

Proof: Follows straightforward from Corollary 8.1.1.

Next we follow Horisberger and Bélanger [HoB] and provide computationally tractable criterion for stability of the polytope

$$\mathcal{A} = \left\{ A = \sum_{i=1}^{k} \alpha_i A_i \; : \; \sum_{i=1}^{k} \alpha_i = 1, \; \alpha_i \geq 0, \; i = 1, \ldots, k \right\}. \tag{8.1.4}$$

To be precise we are seeking the following **sufficient** stability condition: *Determine whether there exists a positive-definite symmetric matrix P such that*

$$A'P + PA \text{ is negative definite for each } A \in \mathcal{A}.$$

It is easy to see that the condition

$$A_i'P + PA_i \text{ is negative definite for each } i = 1, \ldots, k \tag{8.1.5}$$

implies the former stability condition. Hence, in the remainder of the section we assume that A_1 is stable, and focus on (8.1.5). For each $n \times n$ symmetric matrix P, a vector $\mathbf{x} \in \mathbf{R}^n$, and $i = 1, \ldots, k$ let

$$\phi_{\mathbf{x},i}(P) = \mathbf{x}'\,(A_i'P + PA_i)\,\mathbf{x}.$$

Since each $\phi_{\mathbf{x},i}$ is a linear function on the space of symmetric $n \times n$ matrices, the function

$$\phi(P) = \sup\left\{\phi_{\mathbf{x},i}(P) \; : \; |\mathbf{x}| = 1, \; i = 1, \ldots, k\right\}$$

is convex. If

$$\min\left\{\phi(P) \; : \; P = P'\right\} < 0,$$

then the polytope $\mathcal{A}$ is stable. Indeed, let $P = P'$ be such that $\phi(P) < 0$. Then, in particular, $A_1'P + PA_1$ is negative definite. Since A_1 is stable, Corollary 8.1.2 yields $P > 0$. This implies (8.1.5), and completes the proof.

The robust stability problem for a polytope of matrices thus has been reduced to an *unconstrained* convex minimization problem. In the next section we present results of Boyd et al. [BBFE] reducing a simultaneous stabilization problem to a *constrained* convex minimization problem.

8.2. Simultaneous Stabilization

Consider the linear system with state feedback

$$\frac{d}{dt}\mathbf{x}(t) = A(t)\mathbf{x}(t) + B(t)\mathbf{u}(t), \; \mathbf{u}(t) = K\mathbf{x}(t) \tag{8.2.1}$$

where

$$[A(t)\ B(t)] \in \text{conv}\ \{[A_1\ B_1],\dots,[A_k\ B_k]\}\,.$$

Our objective is to construct a matrix K such that there exists a positive definite and symmetric matrix P so that

$$(A_i + B_iK)'P + P(A_i + B_iK) < 0,\ i = 1,\dots,k. \tag{8.2.2}$$

In other words for each vector $\mathbf{x} \in \mathbf{R}^n$

$$\mathbf{x}'\left[(A_i + B_iK)'P + P(A_i + B_iK)\right]\mathbf{x} < 0. \tag{8.2.3}$$

Note that this inequality is *not* linear in (P, K). However with the liner fractional transformation

$$Y = P^{-1},\ \text{and}\ W = KP^{-1}$$

the inequality (8.2.3) becomes

$$\mathbf{x}'[YA_i' + W'B_i' + A_iY + B_iW]\mathbf{x} < 0.$$

Since for each $\mathbf{x}$ the last inequality is linear in (Y, W) the function

$$\psi(Y, W) = \sup\{\mathbf{x}'[YA_i' + W'B_i' + A_iY + B_iW]\mathbf{x}\ :\ |\mathbf{x}| = 1,\ i = 1,\dots,k\}$$

is convex. The desired feedback matrix K exists if and only if

$$\min\{\psi(Y, W)\ :\ Y = Y' > 0\} < 0.$$

Furthermore, any matrix $K = WY^{-1}$, with $Y = Y' > 0$, and $\psi(Y, W) < 0$ stabilizes the systems (8.2.1).

An application of the Linear Matrix Inequalities (LMI) approach (see [BEFB]) reduces design of a stabilizing state feedback for delay systems to a convex constrained minimization problem (see e.g., [Sk]). For the existing methods and fast algorithms for solving this type of optimization problems we refer the reader to [NN].

Bibliography

[A] Ackermann J., Does it suffice to check a subset of multilinear parameters in robustness analysis? IEEE Trans. Autom. Contr., Vol. AC-37, No. 4, pp. 487-488, 1992.

[AB] Agathoklis P., Bruton L., Practical–BIBO stability of n–dimensional discrete systems. IEE Proceedings, Vol. 130, Pt. G, No. 6, pp. 236-242, 1983.

[AG] Adams W.W., Goldstein L.J., Introduction to Number Theory. Prentice Hall, New Jersey, 1976.

[AH] Avellar C.E., Hale J.K., On the zeros of exponential polynomials. Journal of Mathematical Analysis and Applications, Vol. 73, pp. 434-452, 1980.

[AHK] Ackermann J., Hu H.Z., Kaesbauer D., Robustness analysis: a case study. IEEE Trans. Autom. Contr., Vol. AC-35, No. 3, pp. 352-356, 1990.

[ADM] Anagnost J.J., Desoer C.A., Minnichelli R.J., Generalized Nyquist tests for robust stability: frequency domain generalizations of Kharitonov's theorem, pp. 79-96 in Robustness in Identification and Control, Milanese M., Tempo R., Vicino A., (eds.) International Workshop on Robustness in Identification and Control, Turin, Italy, 1988. Plenum Press, 1989.

[AJM] Anderson B.D.O., Jury E.I., Mansour M., On robust Hurwitz polynomials. IEEE Trans. Autom. Contr. Vol. AC-32, No. 10, pp. 909-913, 1987.

[AKMD] Anderson B.D.O., Kraus F., Mansour M., Dasgupta S., Easily testable sufficient conditions for the robust stability of systems with multilinear parameter dependence, pp. 81-92 in Robustness of Dynamic Systems with Parameter Uncertainties. Mansour M., Balemi S., Truol W., (editors). Monte Verità, Birkhäuser Verlag Basel, 1992.

[Bar] Barmish B.R., New Tools for Robustness of Linear Systems. Macmillan, New York, 1994.

[Bart] Bartlett A.C., Counter-example to "Clockwise nature of Nyquist locus of stable transfer functions". Int. J. Contr., Vol. 51, No. 6, pp. 1479-1483, 1990.

[Bas] Basu S., On the multidimensional generalization of robustness of scattering Hurwitz property of complex polynomials. IEEE Trans. on Circuits and Systems, Vol. 36, No. 9, pp. 1159-1167, 1989.

[Ber] Berge C., Topological Spaces. Macmillan, New York, 1963.

[Blo] Blondel V., Simultaneous Stabilization of Linear Systems. Lecture Notes in Control and Information Sciences, Volume 191, Springer-Verlag, London, 1994.

[Boe1] Boese F.G., Stability in a special class of retarded difference-differential equations with interval-valued coefficients. Journal of Mathematical Analysis and Applications, Vol. 181, pp. 227-247, 1994.

[Boe2] Boese F.G., On the stability of real exponential polynomials with interval-valued delays. Multidimensional Systems and Signal Processing, in press.

[Bos1] Bose N.K., Problems and progress in multidimensional systems theory. Proceedings of the IEEE, Vol. 65, No. 6, pp. 824-840, 1977.

[Bos2] Bose N.K., Applied Multidimensional Systems Theory. Van Nostrand Reinhold, 1982.

[Bos3] Bose N.K., Boundary Implication Results in Parameter Space, pp. 47-57 in Handbook of Statistics, Vol. 10. Bose N.K., Rao C.R. (eds.), Elsevier Science Publishers, 1993.

[Bos4] Bose N.K., Inference of Properties of Sets from Subsets, Chapter 4 in Multivariate Analysis: Futire Directions. Rao C.R. (ed.), Elsevier Science Publishers, 1993.

[Bru] Brumley W.E., On the asymptotic behavior of solutions of differential-difference equations of neutral type. Journal of Differential Equations, Vol. 7, pp. 175-188, 1970.

[BB] Boyd S.P., Barratt C.H., Linear Controller Design. Prentice Hall, New Jersey, 1991.

[BC] Bellman R., Cooke K.L., Differential-Difference Equations. Academic Press, New York, 1963.

[BhK] Bhattacharyya S.P., Keel L.H., Robust stability and control of linear and multilinear interval systems, pp. 31-77 in Robust Control Systems, Techniques and Applications. Control and Dynamic Systems, Vol. 51, Part 2, Leondes C.T. (ed.). Academic Press, 1992.

[BoK] Bose N.K., Kim K.D., Stability of a complex polynomial set with coefficients in a diamond and generalizations. IEEE Trans. on Circuits and Systems, Vol. 36, No. 9, pp. 1168-1174, 1989.

[BP] Barmish B.R., Polyak B.T., The volumetric singular value and robustness of feedback control systems. Technical Report ECE–93–9, Department of Electrical and Computer Engineering, University of Wisconsin–Madison, 1993.

[BS1] Barmish B.R., Shi Z., Robust stability of a class of polynomials with coefficients depending multilinearly on perturbations. IEEE Trans. Autom. Contr., Vol. AC-35, No. 9, pp. 1040-1043, 1990.

[BS2] Barmish B.R., Shi Z., Robust stability of perturbed systems with time delays. Automatica, Vol. 25, No. 3, 1989, pp. 371-381.

[BG] Blondel V., Gevers M., The simultaneous stabilizability of three linear systems is rationally undecidable. Math. Contr., Signals Syst., Vol. 6, pp. 135-145, 1994.

[BZ] Bose N.K., Zeheb E., Kharitonov's theorem and stability test for multidimensional digital filters. IEE Proc.–G, Vol. 133, No. 4, pp. 187-190, 1986.

[BAH] Barmish B.R., Ackermann J., Hu H., The tree structured decomposition: A new approach to robust stability analysis. Proceedings of the Conference on Information Sciencies and Systems. Princeton, New Jersy, pp. 133-145, 1990.

[BHH] Bartlett A.C., Hollot C.V., Huang L., Root locations of an entire polytope of polynomials: It suffices to check the edges. Math. Contr., Signals Syst., Vol. 1, pp. 61-71, 1988.

[BBFE] Boyd S., Balakrishnan V., Feron E., El Ghaoui L., Control system analysis and synthesis via linear matrix inequalities. Proceedings of the 1993 ACC, San Francisco, California, pp. 2200-2206, 1993.

[BEFB] Boyd S., El Ghaoui L., Feron E., Balakrishnan V., Linear Matrix Inequalities in Systems and Control Theory. SIAM Studies in Applied Mathematics, Vol. 15, 1994.

[BHKT] Barmish B.R., Hollot C.V., Kraus F.J., Tempo R., Extreme point results for robust stabilization of interval plants with first order compensators. IEEE Trans. Autom. Contr., Vol. AC-37, No. 6, pp. 707-714, 1992.

[BTHK] Barmish B.R., Tempo R., Hollot C.V., Kang H.I., An extreme point result for robust stability of a diamond of polynomials. Proceedings of the 29th IEEE Conference on Decision and Control, Honolulu, Hawaii, pp. 37-38, 1990.

[CB] Chapellat H., Bhattacharyya S.P., A generalization of Kharitonov's theorem: robust stability of interval plants. IEEE Trans. Autom. Contr., Vol. AC-34, No. 3, pp. 306-311, 1989.

[CM] Chui C.K., Mhaskar H.N., On multivariate robust stability. SIAM J. Control and Optimization, Vol. 30, No. 5, pp. 1190-1199, 1992.

[CK1] Cohen N., Kogan J., Convexity of a frequency response arc associated with a stable entire function. Mathematics Research Report KOG 94–03, Department of Mathematics, University of Maryland Baltimore County.

[CK2] Cohen N., Kogan J., Convexity of a frequency response arc associated with a stable quasipolynomial. Mathematics Research Report KOG 94–02, Department of Mathematics, University of Maryland Baltimore County.

[CBD] Chapellat H., Bhattacharyya S.P., Dahleh M., Robust stability of a family of disc polynomials. Int. J. Contr., Vol. 51, pp. 1353-1362, 1990.

[CBL] Chiasson J., Brierley S.D., Lee E.B., A simplified derivation of the Zeheb–Walach 2-D stability test with applications to time–delay systems. IEEE Trans. Autom. Contr., Vol. AC-30, No. 4, pp. 411-414, 1985.

[CFN] Chen J., Fan M.K.H., Nett C.N., On μ and stability of uncertain polynomials. Proceedings of the 1990 ACC, pp. 2200-2206.

[CKB] Chapellat H., Keel L.H., Bhattacharyya S.P., Robustness properties of multilinear interval systems, pp. 73-80 in Robustness of Dynamic Systems with Parameter Uncertainties. Mansour M., Balemi S., Truol W., (editors). Monte Verità, Birkhäuser Verlag Basel, 1992.

[D] Doyle J., Analysis of feedback systems with structured uncertainties. IEE Proceedings, Vol. 129, Part D, pp. 242-250, 1982.

[DH] Djaferis T.E., Hollot C.V., Parameter partitioning via shaping condition for the stability of families of polynomials. IEEE Trans. Autom. Contr., Vol. AC-34, No. 11, pp. 1205-1209, 1989.

[DS] De Gaston R.R.E., Safonov M.G., Exact calculation of the multiloop stability margin. IEEE Trans. Autom. Contr., Vol. AC-33, No. 2, pp. 156-171, 1988.

[DCC] Desages A.C., Castro L., Cendra H., Distance of a complex coefficient stable polynomial from the boundary of the stability set. Multidimensional Systems and Signal Processing, Vol. 2, pp. 189-210, 1991.

[FD] Frazer R.A., Duncan W.J., On the criteria for stability for small motions. Proceedings of the Royal Society A, Vol. 124, pp. 642-654, 1929.

[FS1] Foo Y.K., Soh Y.C., Root clustering of interval polynomials in the left–sector. Systems and Control Letters, Vol. 13, pp. 239-245, 1989.

[FS2] Foo Y.K., Soh Y.C., Stability of a family of polynomials with coefficients bounded in a diamond. IEEE Transactions on Automatic Control, Vol. AC-36, No. 12, pp. 1501-1502, 1991.

[FDB] Fu M., Dasgupta S., Blondel V., Robust stability under a class of nonlinear parametric perturbations. Preprint, 1992.

[FOP1] Fu M., Olbrot A.W., Polis M.P., Robust stability for time–delay systems: the edge theorem and graphical tests. IEEE Transactions on Automatic Control, Vol. 34, No. 8, pp. 813-820, 1989.

[FOP2] Fu M., Olbrot A.W., Polis M.P., The Edge theorem and graphical tests for robust stability of neutral time–delay systems. Automatica, Vol. 27, No. 4, pp. 739-741, 1991.

[Ga] Gantmacher F.R., The Theory of Matrices, Chelsea, New York, 1964.

[Go1] Goodman D., Some stability properties of two–dimensional linear shift–invariant digital filters. IEEE Trans. Circuits Syst., Vol. CAS-24, No. 4, pp. 201-207, 1977.

[Go2] Goodman D., Some difficulties with the double bilinear transformation in 2–D recursive filter design. Proceedings of the IEEE, Vol. 66, No. 7, pp. 796-797, 1978.

[H] Horowitz I., Survey of quantitative feedback theory (QFT). Int. J. Contr., Vol. 53, pp. 255-291, 1991.

[HaB] Hamann J.C., Barmish B.R., Convexity of frequency response arcs associated with a stable polynomial. IEEE Trans. Autom. Contr. Vol. AC-38, No. 6, pp. 904-915, 1993.

[HP1] Hinrichsen D., Pritchard A.J., Stability radius for structured perturbations and the algebraic Riccati equation. Systems and Control Letters, Vol. 8, pp. 105-113, 1986.

[HP2] Hinrichsen D., Pritchard A.J., Real and complex stability radii: a survey. pp. 119-162 in Control of Uncertain Systems, D. Hinrichsen, A.J. Pritchard eds., Proceedings of an International Workshop, Bremen, West Germany, June 1989. Boston, Birkhauser, 1990.

[HP3] Hinrichsen D., Pritchard A.J., Robustness measures for linear systems with application to stabilit radii of Hurwitz and Schur polynomials. Int. J. Contr., Vol. 55, No. 4, pp. 809-844, 1992.

[HS] Holohan A.M., Safonov M.G., Some counterexamples in robust stability theory. Systems and Control Letters, Vol. 21, pp. 95-102, 1993.

[HX] Hollot C.V., Xu Z.L., When the image of a multilinear function is a polytope? A conjecture. Proceedings of the 28th IEEE Conference on Decision and Control, Tampa, Florida, pp. 1890-1891, 1989.

[HoB] Horisberger H.P., Bélanger P.R., Regulators for linear, time invariant plants with uncertain parameters. IEEE Trans. Autom. Contr., Vol. AC-21, No. 5, pp. 705-708, 1976.

[HBA] Horowitz I., Ben–Adam S., Clockwise nature of Nyquist locus of stable transfer function. Int. J. Contr., Vol. 49, pp. 1433-1436, 1989.

[HIT] Hale J.K., Infante E.F., Tsen F.-S.P., Stability in linear delay equations. Journal of Mathematical Analysis and Applications, Vol. 105, pp. 533-555, 1985.

[HKZ1] Hocherman J., Kogan J., Zeheb E., Simple stability criterion for quasipolynomial families with uncertain coefficients and uncertain delays. Proceedings of the 32th IEEE Conference on Decision and Control, San Antonio, Texas, 1993.

[HKZ2] Hocherman J., Kogan J., Zeheb E., On exponential stability of linear systems and Hurwitz stability of characteristic quasipolynomials. EE Publication No. 892, Department of Electrical Engineering, Technion–Israel Institute of Technology, 1993.

[HKKZ] Hocherman J., Kharitonov V.L., Kogan J., Zeheb E., On the stability of quasipolynomials with weighted diamond coefficients. Multidimensional Systems and Signal Processing, in press.

[J1] Jury E.I., Stability of multidimensional scalar and matrix polynomials. Proceedings of the IEEE, Vol. 66, No. 9, pp. 1018-1047, 1978.

[J2] Jury E.I., Inners and Stability of Dynamic Systems. Wiley, New York, 1974.

[J3] Jury E.I., Stability of multidimensional systems and related problems. Chapter 3 in Multidimensional Systems: Techniques and Applications. S.G. Tzafestas (ed.), Marcel Dekker, New York, 1985.

[JB] Jury E.I., Bauer P., On the stability of two–dimensional continuous systems. IEEE Trans. on Circuits and Systems, Vol. 35, No. 12, pp. 1487-1500, 1988.

[Kah] Kahan W., Numerical linear algebra. Canadian Math. Bull., Vol. 9, pp. 756-801, 1966.

[Kai] Kailath T., Linear Systems. Prentice–Hall, Englewood Cliffs, N.J., 1980.

[Kam] Kamen E.W., On the relation between zero criteria for two variable polynomials and asymptotic stability of delay differential equations. IEEE Trans. Autom. Contr., Vol. AC-25, No. 5, pp. 983-984, 1980.

[Kh] Kharitonov V.L., Asymptotic stability of an equilibrium position of a family of systems of linear differential equations. Differential Equations, Vol. 14, pp. 1483-1485, 1979.

[Ko1] Kogan J., Computation of stability radius for families of bivariate polynomials. Multidimensional Systems and Signal Processing, Vol. 4, 1993, pp. 151-165.

[Ko2] Kogan J., Stability of a polynomial and convexity of a frequency response arc. Proceedings of the 32th IEEE Conference on Decision and Control, San Antonio, Texas, 1993.

[KJ1] Katbab A., Jury E.I., Robust Schur stability of a complex coefficient polynomial set with coefficients in a diamond. Journal of the Franklin Institute, Vol. 327, No. 5, pp. 687-698, 1990.

[KJ2] Katbab A., Jury E.I., Generalization and comparison of two recent frequency–domain stability robustness results. Int. J. Contr., Vol. 53, No. 2, pp. 463-475, 1991.

[KL1] Kogan J., Leizarowitz A., Frequency domain criterion for robust stability of interval time–delay systems. Automatica, in press.

[KL2] Kogan J., Leizarowitz A., Exponential stability of linear systems with commensurate time–delays. Mathematics Research Report, March 28, 1994. Department of Mathematics, University of Maryland Baltimore County.

[KM] Kraus F.J., Mansour M., On robust stability of discrete systems. Proceedings of the 29th IEEE Conference on Decision and Control, Honolulu, Hawaii, pp. 421-422, 1990.

[KP] Kiselev O.N., Polyak B.T., Robust gain margin for a cascade of uncertain links. Preprint, 1994.

[KT] Kharitonov V.L., Tempo R., On stability of a weighted diamond of real polynomials CENS-CNR Technical Report 11, 1992.

[KZ] Kharitonov V.L., Zhabko A.P., Robust stability of time–delay systems. IEEE Trans. Autom. Contr., in press.

[KAM] Kraus F.J., Anderson B.D.O., Mansour M., Robust stability of polynomials with multilinear parameter dependence. Int. J. Contr., Vol. 50, No. 5, pp. 1745-1762, 1989.

[Lei] Leitmann G., On one approach to the control of uncertain systems. Transactions of the ASME, Journal of Dynamic Systems, Measurement, and Control, Vol. 115, pp. 373-380, 1993.

[Lev] Levin B.Ja., Distribution of Zeros of Entire Functions. Translations of Mathematical Monographs, Vol. 5, American Mathematical Society. Providence, Rhode Island, 1964.

[Ll] Lloyd N.G., Degree Theory. Cambridge University Press. Cambridge, London, New York, Melbourne, 1978.

[LCZ] Levkovich A., Cohen N., Zeheb E., A root distribution criterion for an interval polynomial in a sector. Preprint, 1994.

[LNL] Li Y., Nagpal K.M., Lee E.B., Stability analysis of polynomials with coefficients in disks. IEEE Trans. Autom. Contr., Vol. AC-37, No. 4, pp. 509-513, 1992.

[M] Martin J.M., State-space measures for stability robustness. IEEE Transactions on Automatic Control, Vol. 32, No. 6, pp. 509-512, 1987.

[MAD] Minnichelli R.J., Anagnost J.J., Desoer C.A., An elementary proof of Kharitonov's stability theorem with extensions. IEEE Trans. Autom. Contr., Vol. AC-34, No. 9, pp. 995-998, 1989.

[MZJ] Malek-Zavarei M., Jamshidi M., Time-Delay Systems. Analysis, Optimization and Applications. North-Holland Systems and Control Series, Vol. 9. North-Holland, Amsterdam, New York, Oxford, Tokyo, 1987.

[Nei] Neimark Y.I., Stability of Linearized Systems (in Russian). Leningrad, LKVVIA, 1949.

[Nem] Nemirovskii A., Several NP-hard problems arising in robust stability analysis. Math. Contr., Signals Syst. Vol. 6, pp. 99-105, 1994.

[NN] Nesterov Yu., Nemirovskii A., Interior-Point Polynomial Algorithms in Convex Programming. SIAM Studies in Applied Mathematics, Vol. 13, 1994.

[P] Pontryagin L.S., On zeros of some elementary transcendental functions. Translations of AMS, Ser. 2, Vol. 1, pp. 95-110, 1955.

[Pe] Petersen I.R., A class of stability regions for which a Kharitonov-like theorem hold. IEEE Trans. Autom. Contr., Vol. AC-34, No. 9, pp. 1111-1115, 1989.

[Po] Polyak B.T., Robustness analysis for multilinear perturbations, pp. 93-104 in Robustness of Dynamic Systems with Parameter Uncertainties. Mansour M., Balemi S., Truol W., (editors). Monte Verità, Birkhäuser Verlag Basel, 1992.

[PD] Packard A., Doyle J., Complex structural singular value. Automatica, Vol. 29, No. 9, pp. 71-110, 1993.

[PK] Polyak B.T., Kogan J., Necessary and sufficient conditions for robust stability of multiaffine systems. Mathematics Research Report 93-06, Department of Mathematics, University of Maryland Baltimore County.

[PL] Polyak B.T., Lan L.H., Value set of transfer functions with parametric uncertainty and their applications in robustness analysis. Preprint, 1993.

[PS] Polya G., Szegö G., Problems and Theorems in Analysis, Vol. I. Springer, Berlin, 1972.

[PT1] Polyak B.T., Tsypkin Ya.Z., Frequency criteria of robust stability and aperiodicity of linear systems. Automation and Remote Control, Vol. 51, No. 9, Part 1, pp. 1192-1201, 1990.

[PT2] Polyak B.T., Tsypkin Ya.Z., Robust stability under complex perturbations of parameters. Automation and Remote Control, Vol. 52, No. 8, Part 1, pp. 1069-1077, 1991.

[PT3] Polyak B.T., Tsypkin Ya.Z., Robust stability of the linear discrete systems. Dokl. Acad. Nauk, Vol. 316, No. 4, pp. 842-845, 1991, (in Russian).

[PDB] Packard A., Doyle J., Balas G., Linear, multivariate robust control with a μ perspective. Transactions of the ASME, Journal of Dynamic Systems, Measurement, and Control, Vol. 115, pp. 426-438, 1993.

[QD] Qiu L., Davison E.J., A unified approach for the stability robustness of polynomials in a convex set. Automatica, Vol. 28, No. 5, pp. 945-959, 1992.

[QBRDYD] Qiu L., Bernhardsson B., Rantzer A., Davison E.J., Young P.M., Doyle J.C., A formula for computation of the real stability radius. Preprint, 1993.

[Ra1] Rantzer A., Kharitonov's weak theorem holds if and only if the stability region and its reciprocal are convex. Int. J. of Nonlinear and Robust Contr., Vol. 3, pp. 55-62, 1993.

[Ra2] Rantzer A., Stability conditions for polytopes of polynomials. IEEE Trans. Autom. Contr., Vol. AC-37, No. 1, pp. 79-89, 1992.

[Ro] Rockafellar R.T., Convex Analysis. Princeton University Press, 1970.

[Ru] Rudin W., Real and Complex Analysis. McGraw-Hill, New York, 1987.

[RJ] Reddy H.C., Jury E.I., Study of the BIBO stability of 2-D recursive digital filters in the presence of nonessential singularities of the second kind-analog approach. IEEE Trans. Circuits Syst., Vol. CAS-34, No. 3, pp. 280-284, 1987.

[RDC] Robledo C., Desages A., Cendra H., On the distance to the instability border of Hurwitz polynomials that depend affinely on m parameters and a particular convexity property of H^n. Multidimensional Systems and Signal Processing, Vol. 3, No. 1, pp. 45-62, 1992.

[Š] Šiljak D., Nonlinear Systems, the Parameter Analysis and Design. Wiley, New York, 1969.

[Sa] Saeki M., A method of robust stability analysis with highly structured uncertainties. IEEE Trans. Autom. Contr., Vol. AC-31, No. 10, pp. 935-940, 1986.

[Sc] Schwengeler E., Geometrisches über die Verteilung der Nullstellen spezieller ganzer Funktionen (Exponentialsummen). Dissertation, Zurich, 1925.

[Sh] Shcherbakov P.S., Alexander Mikhailovitch Lyapunov: On the centenary of his doctoral dissertation on stability of motion. Automatica, Vol. 28, No. 5, pp. 865-871, 1992.

[Si] Silkowski R., A star-shaped condition for stability of linear retarded functional differential equations. Proceedings of the Royal Society of Edinburgh, Vol. 83A, pp. 189-198, 1979.

[Sk] Skorodinskii V.I., Iterational method of construction of Lyapunov-Krasovskii functionals for linear systems with delay. Automation and Remote Control, Vol. 51, No. 9, pp. 1205-1212, 1990.

[St] Stépán G., Retarded Dynamical Systems: Stability and Characteristic Functions. π Pitman Research Notes in Mathematics Series: 210, 1989.

[SZ] Shi Y.D., Zhou S.F., Stability of a set of multivariate complex polynomials with coefficients varying in a diamond domain. IEEE Trans. Circuits Syst., Vol. CAS-39, No. 8, pp. 683-688, 1992.

[SEP1] Soh Y.C., Evans R.J., Petersen I.R., Characterization of a family of polynomials with interval roots. Technical Report EE8543, University of Newcastle, Newcastle, Australia, 1985.

[SEP2] Soh Y.C., Evans R.J., Petersen I.R., A class of polynomials with multilinear parameter perturbations. Preprint, 1992.

[SRP] Swamy M.N., Roytman L.M., Plotkin E.I., Planar least squares inverse polynomials and practical–BIBO stabilization of n–dimensional linear shift–invariant filters. IEEE Trans. Circuits Syst., Vol. CAS-32, No. 12, pp. 1255-1260, 1985.

[SEPB] Soh Y.C., Evans R.J., Petersen I.R., Betz R.E., Robust pole assignment. Automatica, Vol. 23, No. 5, pp. 601-610, 1987.

[Te] Tempo R., A dual result to Kharitonov's theorem. IEEE Trans. Autom. Contr., Vol. AC-35, No. 2, pp. 195-198, 1990.

[Ti] Titchmarsh E.C., The Theory of Functions. Oxford University Press, London, 1962.

[TF] Tsypkin Ya.Z., Fu M., Robust stability of time–delay systems with an uncertain time–delay constant. Int. J. Contr., Vol. 57, No. 4, pp. 865-879, 1993.

[TK] Teboulle M., Kogan J., Applications of optimization methods to robust stability of linear systems. Journal of Optimization Theory and Applications, Vol. 81, No. 1, 1994, pp. 169-192.

[TP1] Tsypkin Ya.Z., Polyak B.T., Frequency domain approach to robust stability of continuous systems, pp. 389-399 in Systems and Control: Topics in Theory and Applications. Kozin F., Ono T., (eds.), MITA Press, Osaka, 1991.

[TP2] Tsypkin Ya.Z., Polyak B.T., Frequency domain criteria for l^p–robust stability of continuous linear systems. IEEE Trans. Autom. Contr., Vol. AC-36, No. 12, pp. 1464-1469, 1991.

[TT] Tsing N.-K., Tits A., When is a multiaffine image of a cube a polygon? Systems and Control Letters, Vol. 20, pp. 439-445, 1993.

[TVZ] Tesi A., Vicino A., Zappa G., Clockwise property of the Nyquist plot with implications for absolute stability. Automatica, Vol. 28, No. 1, pp. 71-80, 1992.

[WF] Wang S., Fairman F.W., On the simultaneous robust stabilization of three plants. Int. J. Contr., Vol. 59, No. 2, pp. 1095-1106, 1994.

[WH] Wang L., Huang L., Robust stability of diamond families of polynomials with complex coefficients. Int. J. Systems Sci., Vol. 3, No. 8, pp. 1371-1378, 1992.

[XF] Xin X., Feng C.-B., Robust stability of control systems with parametric uncertainties. Proceedings of the 31th IEEE Conference on Decision and Control, Tucson, Arizona, pp. 1559-1562, 1992.

[Z] Zeheb E., Necessary and sufficient conditions for robust stability of a continuous systems–the continuous dependency case illustrated via multilinear dependency. IEEE Trans. on Circuits and Systems, Vol. 37, No. 1, pp. 47-53, 1990.

[ZD] Zadeh L., Desoer C.A., Linear System Theory. McGraw Hill, NY, 1963.

[ZW] Zeheb E., Walach E., Zero sets of multiparameter functions and stability of multidimensional systems. IEEE Trans. Acous., Speech, Sig. Processing, ASSP–29, pp. 197-206, 1981.

List of Special Symbols

Author Index

Subject Index

Lecture Notes in Control and Information Sciences

Edited by M. Thoma

1989–1994 Published Titles:

Vol. 135: Nijmeijer, Hendrik; Schumacher, Johannes M. (Eds)
Three Decades of Mathematical System Theory. A Collection of Surveys at the Occasion of the 50th Birthday of Jan C. Willems.
562 pp. 1989 [3-540-51605-0]

Vol. 136: Zabczyk, Jerzy W. (Ed.)
Stochastic Systems and Optimization. Proceedings of the 6th IFIP WG 7.1 Working Conference, Warsaw, Poland, September 12-16, 1988.
374 pp. 1989 [3-540-51619-0]

Vol. 137: Shah, Sirish L.; Dumont, Guy (Eds)
Adaptive Control Strategies for Industrial Use. Proceedings of a Workshop held in Kananaskis, Canada, 1988.
360 pp. 1989 [3-540-51869-X]

Vol. 138: McFarlane, Duncan C.; Glover, Keith
Robust Controller Design Using Normalized Coprime Factor Plant Descriptions.
206 pp. 1990 [3-540-51851-7]

Vol. 139: Hayward, Vincent; Khatib, Oussama (Eds)
Experimental Robotics I. The First International Symposium, Montreal, June 19-21, 1989.
613 pp. 1990 [3-540-52182-8]

Vol. 140: Gajic, Zoran; Petkovski, Djordjija; Shen, Xuemin (Eds)
Singularly Perturbed and Weakly Coupled Linear Control Systems. A Recursive Approach.
202 pp. 1990 [3-540-52333-2]

Vol. 141: Gutman, Shaul
Root Clustering in Parameter Space.
153 pp. 1990 [3-540-52361-8]

Vol. 142: Gündes, A. Nazli; Desoer, Charles A.
Algebraic Theory of Linear Feedback Systems with Full and Decentralized Compensators.
176 pp. 1990 [3-540-52476-2]

Vol. 143: Sebastian, H.-J.; Tammer, K. (Eds)
System Modelling and Optimizaton. Proceedings of the 14th IFIP Conference, Leipzig, GDR, July 3-7, 1989.
960 pp. 1990 [3-540-52659-5]

Vol. 144: Bensoussan, A.; Lions, J.L. (Eds)
Analysis and Optimization of Systems. Proceedings of the 9th International Conference. Antibes, June 12-15, 1990.
992 pp. 1990 [3-540-52630-7]

Vol. 145: Subrahmanyam, M. Bala
Optimal Control with a Worst-Case Performance Criterion and Applications.
133 pp. 1990 [3-540-52822-9]

Vol. 146: Mustafa, Denis; Glover, Keith
Minimum Enthropy H Control.
144 pp. 1990 [3-540-52947-0]

Vol. 147: Zolesio, J.P. (Ed.)
Stabilization of Flexible Structures. Third Working Conference, Montpellier, France, January 1989.
327 pp. 1991 [3-540-53161-0]

Vol. 148: Not published

Vol. 149: Hoffmann, Karl H; Krabs, Werner (Eds)
Optimal Control of Partial Differential Equations. Proceedings of IFIP WG 7.2 - International Conference. Irsee, April, 9-12, 1990.
245 pp. 1991 [3-540-53591-8]

Vol. 150: Habets, Luc C.
Robust Stabilization in the Gap-topology.
126 pp. 1991 [3-540-53466-0]

Vol. 151: Skowronski, J.M.; Flashner, H.; Guttalu, R.S. (Eds)
Mechanics and Control. Proceedings of the 3rd Workshop on Control Mechanics, in Honor of the 65th Birthday of George Leitmann, January 22-24, 1990, University of Southern California.
497 pp. 1991 [3-540-53517-9]

Vol. 152: Aplevich, J. Dwight
Implicit Linear Systems.
176 pp. 1991 [3-540-53537-3]

Vol. 153: Hajek, Otomar
Control Theory in the Plane.
269 pp. 1991 [3-540-53553-5]

Vol. 154: Kurzhanski, Alexander; Laseicka, Irena (Eds)
Modelling and Inverse Problems of Control for Distributed Parameter Systems. Proceedings of IFIP WG 7.2 - IIASA Conference, Laxenburg, Austria, July 1989.
170 pp. 1991 [3-540-53583-7]

Vol. 155: Bouvet, Michel; Bienvenu, Georges (Eds)
High-Resolution Methods in Underwater Acoustics.
244 pp. 1991 [3-540-53716-3]

Vol. 156: Hämäläinen, Raimo P.; Ehtamo, Harri K. (Eds)
Differential Games - Developments in Modelling and Computation. Proceedings of the Fourth International Symposium on Differential Games and Applications, August 9-10, 1990, Helsinki University of Technology, Finland.
292 pp. 1991 [3-540-53787-2]

Vol. 157: Hämäläinen, Raimo P.; Ehtamo, Harri K. (Eds)
Dynamic Games in Economic Analysis. Proceedings of the Fourth International Symposium on Differential Games and Applications. August 9-10, 1990, Helsinki University of Technology, Finland.
311 pp. 1991 [3-540-53785-6]

Vol. 158: Warwick, Kevin; Karny, Miroslav; Halouskova, Alena (Eds)
Advanced Methods in Adaptive Control for Industrial Applications.
331 pp. 1991 [3-540-53835-6]

Vol. 159: Li, Xunjing; Yong, Jiongmin (Eds)
Control Theory of Distributed Parameter Systems and Applications. Proceedings of the IFIP WG 7.2 Working Conference, Shanghai, China, May 6-9, 1990.
219 pp. 1991 [3-540-53894-1]

Vol. 160: Kokotovic, Petar V. (Ed.)
Foundations of Adaptive Control.
525 pp. 1991 [3-540-54020-2]

Vol. 161: Gerencser, L.; Caines, P.E. (Eds)
Topics in Stochastic Systems: Modelling, Estimation and Adaptive Control.
1991 [3-540-54133-0]

Vol. 162: Canudas de Wit, C. (Ed.)
Advanced Robot Control. Proceedings of the International Workshop on Nonlinear and Adaptive Control: Issues in Robotics, Grenoble, France, November 21-23, 1990.
Approx. 330 pp. 1991 [3-540-54169-1]

Vol. 163: Mehrmann, Volker L.
The Autonomous Linear Quadratic Control Problem. Theory and Numerical Solution.
177 pp. 1991 [3-540-54170-5]

Vol. 164: Lasiecka, Irena; Triggiani, Roberto
Differential and Algebraic Riccati Equations with Application to Boundary/Point Control Problems: Continuous Theory and Approximation Theory.
160 pp. 1991 [3-540-54339-2]

Vol. 165: Jacob, Gerard; Lamnabhi-Lagarrigue, F. (Eds)
Algebraic Computing in Control. Proceedings of the First European Conference, Paris, March 13-15, 1991.
384 pp. 1991 [3-540-54408-9]

Vol. 166: Wegen, Leonardus L. van der
Local Disturbance Decoupling with Stability for Nonlinear Systems.
135 pp. 1991 [3-540-54543-3]

Vol. 167: Rao, Ming
Integrated System for Intelligent Control.
133 pp. 1992 [3-540-54913-7]

Vol. 168: Dorato, Peter; Fortuna, Luigi; Muscato, Giovanni
Robust Control for Unstructured Perturbations: An Introduction.
118 pp. 1992 [3-540-54920-X]

Vol. 169: Kuntzevich, Vsevolod M.; Lychak, Michael
Guaranteed Estimates, Adaptation and Robustness in Control Systems.
209 pp. 1992 [3-540-54925-0]

Vol. 170: Skowronski, Janislaw M.; Flashner, Henryk; Guttalu, Ramesh S. (Eds)
Mechanics and Control. Proceedings of the 4th Workshop on Control Mechanics, January 21-23, 1991, University of Southern California, USA.
302 pp. 1992 [3-540-54954-4]

Vol. 171: Stefanidis, P.; Paplinski, A.P.; Gibbard, M.J.
Numerical Operations with Polynomial Matrices: Application to Multi-Variable Dynamic Compensator Design.
206 pp. 1992 [3-540-54992-7]

Vol. 172: Tolle, H.; Ersü, E.
Neurocontrol: Learning Control Systems Inspired by Neuronal Architectures and Human Problem Solving Strategies.
220 pp. 1992 [3-540-55057-7]

Vol. 173: Krabs, W.
On Moment Theory and Controllability of Non-Dimensional Vibrating Systems and Heating Processes.
174 pp. 1992 [3-540-55102-6]

Vol. 174: Beulens, A.J. (Ed.)
Optimization-Based Computer-Aided Modelling and Design. Proceedings of the First Working Conference of the New IFIP TC 7.6 Working Group, The Hague, The Netherlands, 1991.
268 pp. 1992 [3-540-55135-2]

Vol. 175: Rogers, E.T.A.; Owens, D.H.
Stability Analysis for Linear Repetitive Processes.
197 pp. 1992 [3-540-55264-2]

Vol. 176: Rozovskii, B.L.; Sowers, R.B. (Eds)
Stochastic Partial Differential Equations and their Applications. Proceedings of IFIP WG 7.1 International Conference, June 6-8, 1991, University of North Carolina at Charlotte, USA.
251 pp. 1992 [3-540-55292-8]

Vol. 177: Karatzas, I.; Ocone, D. (Eds)
Applied Stochastic Analysis. Proceedings of a US-French Workshop, Rutgers University, New Brunswick, N.J., April 29-May 2, 1991.
317 pp. 1992 [3-540-55296-0]

Vol. 178: Zolésio, J.P. (Ed.)
Boundary Control and Boundary Variation. Proceedings of IFIP WG 7.2 Conference, Sophia- Antipolis,France, October 15-17, 1990.
392 pp. 1992 [3-540-55351-7]

Vol. 179: Jiang, Z.H.; Schaufelberger, W.
Block Pulse Functions and Their Applications in Control Systems.
237 pp. 1992 [3-540-55369-X]

Vol. 180: Kall, P. (Ed.)
System Modelling and Optimization. Proceedings of the 15th IFIP Conference, Zurich, Switzerland, September 2-6, 1991.
969 pp. 1992 [3-540-55577-3]

Vol. 181: Drane, C.R.
Positioning Systems - A Unified Approach.
168 pp. 1992 [3-540-55850-0]

Vol. 182: Hagenauer, J. (Ed.)
Advanced Methods for Satellite and Deep Space Communications. Proceedings of an International Seminar Organized by Deutsche Forschungsanstalt für Luft-und Raumfahrt (DLR), Bonn, Germany, September 1992.
196 pp. 1992 [3-540-55851-9]

Vol. 183: Hosoe, S. (Ed.)
Robust Control. Proceesings of a Workshop held in Tokyo, Japan, June 23-24, 1991.
225 pp. 1992 [3-540-55961-2]

Vol. 184: Duncan, T.E.; Pasik-Duncan, B. (Eds)
Stochastic Theory and Adaptive Control. Proceedings of a Workshop held in Lawrence, Kansas, September 26-28, 1991.
500 pages. 1992 [3-540-55962-0]

Vol. 185: Curtain, R.F. (Ed.); Bensoussan, A.; Lions, J.L.(Honorary Eds)
Analysis and Optimization of Systems: State and Frequency Domain Approaches for Infinite-Dimensional Systems. Proceedings of the 10th International Conference, Sophia-Antipolis, France, June 9-12, 1992.
648 pp. 1993 [3-540-56155-2]

Vol. 186: Sreenath, N.
Systems Representation of Global Climate Change Models. Foundation for a Systems Science Approach.
288 pp. 1993 [3-540-19824-5]

Vol. 187: Morecki, A.; Bianchi, G.; Jaworeck, K. (Eds)
RoManSy 9: Proceedings of the Ninth CISM-IFToMM Symposium on Theory and Practice of Robots and Manipulators.
476 pp. 1993 [3-540-19834-2]

Vol. 188: Naidu, D. Subbaram
Aeroassisted Orbital Transfer: Guidance and Control Strategies.
192 pp. 1993 [3-540-19819-9]

Vol. 189: Ilchmann, A.
Non-Identifier-Based High-Gain Adaptive Control.
220 pp. 1993 [3-540-19845-8]

Vol. 190: Chatila, R.; Hirzinger, G. (Eds)
Experimental Robotics II: The 2nd International Symposium, Toulouse, France, June 25-27 1991.
580 pp. 1993 [3-540-19851-2]

Vol. 191: Blondel, V.
Simultaneous Stabilization of Linear Systems.
212 pp. 1993 [3-540-19862-8]

Vol. 192: Smith, R.S.; Dahleh, M. (Eds)
The Modeling of Uncertainty in Control Systems.
412 pp. 1993 [3-540-19870-9]

Vol. 193: Zinober, A.S.I. (Ed.)
Variable Structure and Lyapunov Control
428 pp. 1993 [3-540-19869-5]

Vol. 194: Cao, Xi-Ren
Realization Probabilities: The Dynamics of Queuing Systems
336 pp. 1993 [3-540-19872-5]

Vol. 195: Liu, D.; Michel, A.N.
Dynamical Systems with Saturation Nonlinearities: Analysis and Design
212 pp. 1994 [3-540-19888-1]

Vol. 196: Battilotti, S.
Noninteracting Control with Stability for Nonlinear Systems
196 pp. 1994 [3-540-19891-1]

Vol. 197: Henry, J.; Yvon, J.P. (Eds)
System Modelling and Optimization
975 pp approx. 1994 [3-540-19893-8]

Vol. 198: Winter, H.; Nüßer, H.-G. (Eds)
Advanced Technologies for Air Traffic Flow Management
225 pp approx. 1994 [3-540-19895-4]

Vol. 199: Cohen, G.; Quadrat, J.-P. (Eds)
11th International Conference on Analysis and Optimization of Systems – Discrete Event Systems: Sophia-Antipolis, June 15–16–17, 1994
648 pp. 1994 [3-540-19896-2]

Vol. 200: Yoshikawa, T.; Miyazaki, F. (Eds)
Experimental Robotics III: The 3rd International Symposium, Kyoto, Japan, October 28-30, 1993
624 pp. 1994 [3-540-19905-5]